Agriculture and Food

Peter Atkins

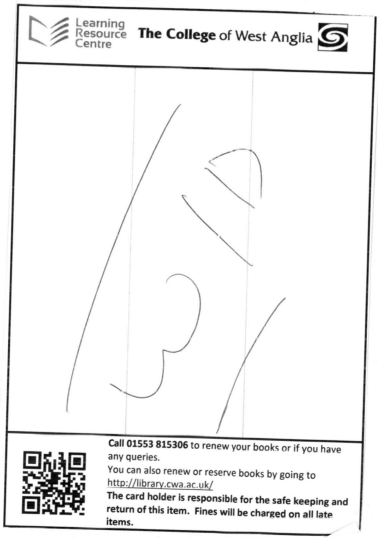

Coll... Educational

An imp...

Contents

Skills matrix

Chapter	Understanding of text/ newspaper extracts and classification	Graphical/mapping methods and annotated diagrams	Analysis of data from tables, graphs and diagrams	Analysis of photographs	Analysis of maps	Statistical analysis/methods	Values enquiry	Project work based on library/fieldwork research/ data collection	Writing: essays, reports, speeches, creative
1	5	2, 4, 6, 7	3	1					
2	1, 2, 4	1, 5	1, 3	11, 13	7, 8, 9 12, 13, 14	10		5	4, 6
3	3, 7, 19	14, 19	1, 4, 5, 6 7, 8, 10, 11 12, 15, 18	2	16, 17		13, 20	3	9, 13, 21
4		12, 14	3, 4, 7, 8, 11, 13, 14, 15, 17, 19	1, 8, 12, 19	2, 5, 6, 9, 10, 12, 16, 18, 19, 20				20
5	1, 7	7	1, 2, 3, 4	1, 3	5, 6		10		8, 9, 11
6	14, 16, 18	2, 3, 11, 13, 14, 17	4, 6, 12, 18	1, 2, 5	2, 3, 7, 8, 9, 15	10			19
7	5, 12	2, 9, 16	1, 2, 3, 6, 7, 8, 11, 12, 16, 17, 18		4, 10, 19	2	14, 15	15	6, 13, 20
8	12, 14	6	3, 5, 8, 9, 13, 15, 16	3	1, 2, 7, 8, 11, 17	4	18	4	10, 18
9	14	1, 3, 9	2, 4, 6, 7, 10	11	8, 13, 15, 16	12	5		4, 5, 17 18
10	7, 10	2	3, 4, 5, 13, 14, 15, 16, 18		6, 17		1, 8, 9	1, 9, 11	8, 12, 19
11		4, 5, 6	4, 10, 12	1	3, 8	5	13	2	7, 9, 11
12	6, 10, 11 14, 17, 18	2, 3, 9	1, 4, 7	8, 12, 15	13, 16		5, 14		18

To the student

The aim of this book is to give you a basic grounding in the important topic of agriculture and food. Both authors are geographers, but the scope of the text is broad enough to be of interest to courses in environmental studies and related disciplines. An important aspect of agriculture and the food system is its integrated nature, which requires a variety of different, interdisciplinary perspectives.

We introduce the concept of the food system by suggesting that you consider food as a commodity passing from its point of production to the act of consumption. This system is made up of a number of stages, such as farming, food production and the distribution chain of wholesaling and retailing. Each stage is influenced by factors ranging from the physical environment and the economic marketplace, to government policies.

To help you in your studies, the book has a number of standard features. Each chapter starts with a map, or plan. This sets out the main themes and the linkages between them. It represents a skeleton on which you build the detailed information supplied in each section.

The text includes numerous tasks which have been devised to help you in three ways. First, there are tests to assess your understanding of the concepts introduced. Second, there are exercises to encourage you to think further, beyond the limited number of ideas and facts we have been able to cover in the space allowed. And third, some tasks will assist you in the development of specific skills (see the skills matrix on page 4).

A complex book of this sort inevitably contains some technical terms, although we have kept these to a minimum. Where further elaboration of a word is required, this is indicated by marking the word in bold each time it first appears in any chapter. Its definition is then included in the glossary at the back of the book.

Finally, each chapter ends with a summary of its main points which are intended to help you recap the material covered. When revision-time comes, you can trace the key ideas in the chapter summaries to the text and relate them to the appropriate case studies and examples.

The appendices provide some statistical materials which should be useful for those attempting coursework and investigative studies. They include explanations of two essential tests: Spearman rank correlation and chi-squared. In addition, some references are supplied to help you with project work.

Peter Atkins

Michael Raw

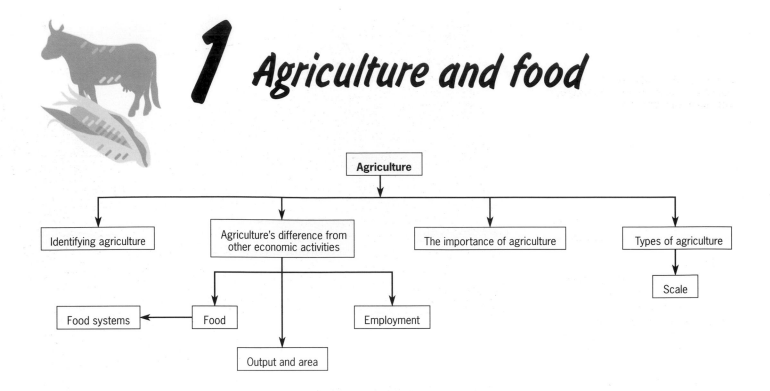

1 Agriculture and food

1.1 Introduction

This book is about agriculture (Chapters 2 to 7), and its principal **output**, food (Chapters 8 to 12). Our aim is to encourage you to think of agriculture and food together, as part of a **food system**. Although agriculture dominates the food system, it is merely the starting point of a complex **food chain**. In economically developed countries the food chain includes wholesalers, food manufacturers, retailers and consumers.

As geographers, we study agriculture and food from the perspective of spatial patterns, spatial processes, people–environment relationships, change and its impact, and the issues which arise from change (Fig. 1.1). However, a complicating factor in all geographical studies is scale. This means that every theme in Figure 1.1 can be studied at a different scale, ranging from individual farms to regions, countries and the world.

International scale (e.g.EC)
National scale (e.g. UK)
Regional scale (e.g. East Anglia)

Spatial patterns	Spatial distribution of wheat cultivation.
Processes/people–environment interaction	Physical conditions needed for wheat growing. Economic environment and government policies.
Change	Expansion of cultivated area and increased intensity of wheat growing.
The impact of processes, interaction and change	Surplus and food mountains; economic and political impact. Environmental impact on landscapes and wildlife.
Issues	Policy of set-aside. Nitrate pollution of water resources. Destruction of hedgerows. Drainage of wetlands.

Figure 1.1 Wheat growing: a geographical perspective

Figure 1.2 A variable rural landscape showing evidence of recent change, Deepdale, Cumbria, UK.

?

1 Study Figure 1.2.
a Describe the spatial patterns of crops, fields and farms.
b From the evidence of the photograph suggest possible processes that might explain the spatial patterns. State whether any of these processes are linked with the physical environment.
c What evidence can you see of changing land-use in the photograph?

2 Draw a diagram like Figure 1.1 to illustrate a geographer's view of food shortages at continental, national and regional scales. You will need to search the index and parts of Chapters 10 to 12. Here are a few ideas to help you: distribution of food shortage, causes of food shortage, famine, malnutrition/undernutrition, out-migration, food aid, food availability.

1.2 What is agriculture?

We all have some idea what agriculture is, and yet it is not so easy to give a precise definition. None the less, for our purposes we have decided on the following: 'agriculture is the control and use of plants and animals for the production of food, fibre and raw materials for industry'. In fact we shall concentrate exclusively on land-based food producing systems. This means excluding forestry, which produces timber and wood pulp for industry, and fishing, which is more often a form of hunting than control.

We need to recognise that agriculture has other purposes apart from producing food and materials (Fig. 1.3). It provides employment for farmers and farm workers; subsistence for many farmers in economically developing countries; recreation and leisure facilities in many economically developed countries; and at the national level it provides **food security**, food exports, and savings on food imports. In addition, a successful farming industry is crucial in preventing the depopulation of rural areas.

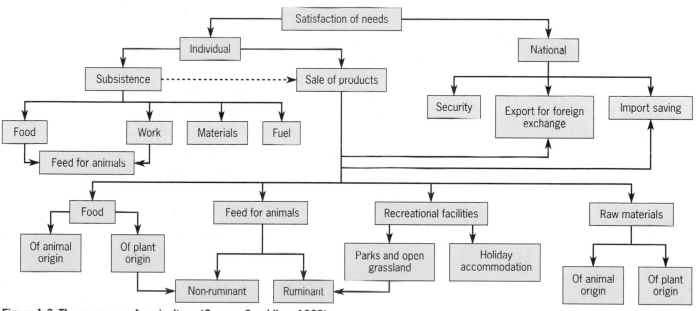

Figure 1.3 The purposes of agriculture (*Source:* Spedding, 1988)

1.3 What makes agriculture different from other economic activities?

Agriculture is different from other economic activities, such as manufacturing and services, in several respects. The main differences are:

1 Agriculture uses large amounts of land. In fact, about one-third of the world's land surface is farmed. Its resources (land, soil, climate etc.) are immobile and area-based which ties agriculture (unlike manufacturing) to specific locations.

2 Agriculture is more strongly influenced by the physical environment than any other major economic activity (Chapters 4 and 5). Although modern technology has made some progress in reducing this dependence, environmental factors remain important.

3 Because agriculture makes use of such a large proportion of the world's land area, its overall environmental impact is greater than other economic activities (Chapter 6). Agriculture creates its own landscapes and has far-reaching effects on the living and non-living components of ecosystems.

4 Agriculture depends on biological processes (Chapter 3). Food products are perishable, and they are produced according to the natural rhythms of the seasons and animal life cycles. This dependence on natural processes inevitably gives a time-lag between investment and yield which might not be acceptable in manufacturing.

5 Most farmers are engaged in several enterprises (Chapter 2), while in contrast, factories are more specialised. However, specialisation is increasingly a feature of modern commercial agriculture, for like manufacturing, it seeks to gain the advantages of **scale economies** (Chapter 7).

6 Small production units (e.g. family farms), with limited turnover and employment, dominate world agriculture (Chapter 7). As a result, individual farmers have little influence on market conditions and prices.

7 Agriculture is more individualistic and less corporate than industry (Chapter 7). Although **agribusiness** is increasingly important in economically developed countries, decision-making by individual farmers is still responsible for most food production.

8 In agriculture, location is fixed and decision-making concentrates on the choice of enterprise (Chapter 7).

9 A large proportion of farm production (especially in the economically developing world) is consumed on the farm and so does not enter trade and other **marketing channels**.

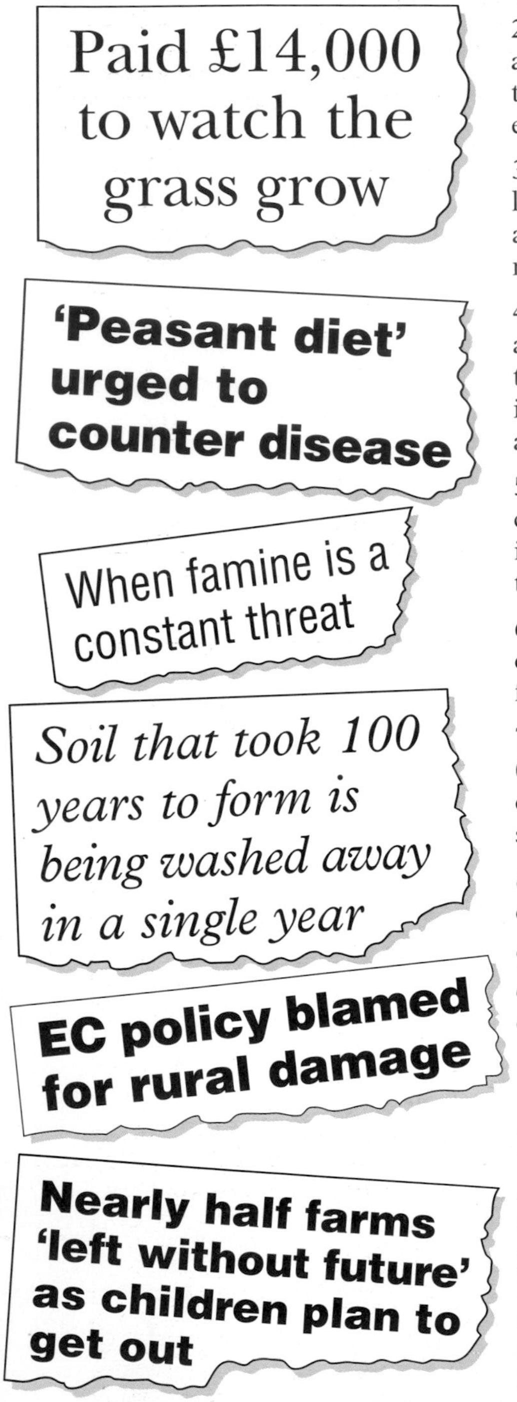

1.4 Why is agriculture important?

Agriculture is, most obviously, the principal source of food for people. It is an activity which provides direct employment for hundreds of millions of people; and if we regard agriculture as the starting point in the food chain, it indirectly creates millions of jobs in related activities such as food manufacturing and food retailing. Agriculture also contributes to economic output and wealth, and as an economic activity, it is by far the largest user of land. Finally, agriculture (and food) are important because they raise issues which affect us all and are given frequent and prominent coverage by the media (Fig. 1.4).

Paid £14,000 to watch the grass grow

'Peasant diet' urged to counter disease

When famine is a constant threat

Soil that took 100 years to form is being washed away in a single year

EC policy blamed for rural damage

Nearly half farms 'left without future' as children plan to get out

Figure 1.4 Newspaper headlines on food issues

Employment

Although, worldwide, employment in agriculture continues to decline, in 1990 agriculture still employed over one billion people, or nearly one in every two economically active people in the world. If we take into account the whole of the food chain, the proportion is even bigger.

Globally, there is considerable spatial unevenness in agricultural employment (Fig. 1.5 and Table 1.1). Thus 95 per cent of the world's agricultural workforce is in economically developing countries, even though these countries only account for 59 per cent of all agricultural land.

Table 1.1 The world's agricultural workforce, 1990 (*Source*: Grigg, 1992)

Region	Agricultural workforce (millions)	% world total	% world population
Africa	136	12.5	9.7
Middle East	35	3.2	5.2
South and South-East Asia	339	31.2	29.0
Asian centrally planned economies	477	44.0	23.5
Latin America	41	3.8	8.4
Other economically developing	1.5	0.1	0.1
All economically developing	1029.5	95.0	75.9
North America	3.5	0.32	5.3
Western Europe	13	1.2	7.4
Australasia	0.55	–	0.4
Other economically developed	6.33	0.58	3.2
Eastern Europe and CIS	31	2.9	7.8
All developed	54.4	5.0	24.1
World total	1083.9	100.0	100.0

?

3 Study Table 1.1 and Figure 1.5.

a Compared to the size of the population, in which regions is the agricultural workforce:
• much larger, • much smaller, than you might expect?

b Suggest possible reasons for the difference between Asian **centrally planned economies** (i.e. mainly China) and North America in Figure 1.6.

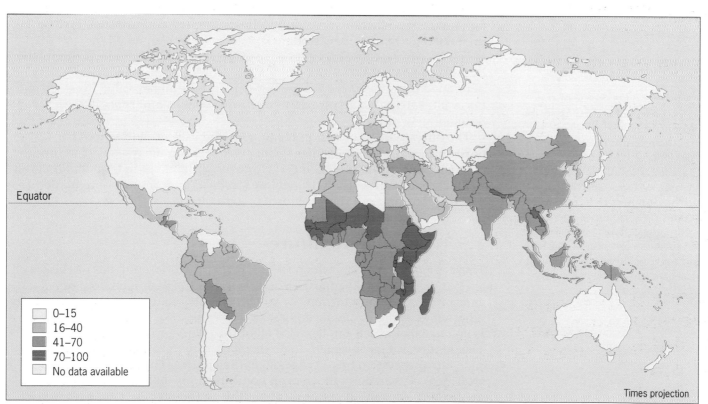

Figure 1.5 Agricultural population as a percentage of the total world population

0–15
16–40
41–70
70–100
No data available

Equator

Times projection

Table 1.2 Percentage of world AGDP and agricultural land area by region, 1990 (*Source*: Grigg, 1992)

Region	World AGDP (%)	World agricultural area (%)
Africa	5.5	17.1
Middle East	4.3	7.5
South and South-East Asia	16.1	6.6
Asian centrally planned economies	12.5	11.8
Latin America	8.4	16.0
North America	11.4	10.8
Western Europe	19.4	3.5
Australasia	1.4	10.4
Other economically developed	9.1	2.1
Eastern Europe and CIS	11.8	14.2

Figure 1.6 The importance of agricultural employment by region

4a Using the information in Tables 1.1 and 1.2 plot two graphs similar to Figure 1.6 of: • AGDP against agricultural land area, • AGDP against the per cent of agricultural workforce.
b What do the graphs tell you about output in relation to land area and workforce in: South and South-East Asia, Asian centrally planned economies, Africa, Western Europe and North America? Do any of your findings surprise you? Explain why.

5 Table 1.3 lists some of the criteria we could use to describe farming systems. Use the index to find the relevant information in this book, and identify the main characteristics of the following farming systems: **shifting cultivation**, wet-rice cultivation, plantation agriculture and agribusiness. The characteristics of hill sheep farming in the UK, and maize cultivation in Iowa are already supplied in Table 1.3.

Agricultural output and agricultural area

Two alternative measures of agriculture's importance are the value of agricultural output (agricultural gross domestic product or AGDP) and agricultural land area (Table 1.2). Again, there are significant regional contrasts in these measures. For example, the value of output in Western Europe is extremely high, even though its agricultural land area is only a tiny fraction of the world total. At the other extreme, Africa's AGDP is insignificant compared to its vast agricultural land area.

However AGDP is not a precise measure of agricultural output. Agriculture in much of the developing world is geared towards **subsistence**. This output, which is consumed on farms, is not therefore included in AGDP. As a result, agricultural production in economically developing countries tends to be understated. On the other hand, AGDP is probably overstated in economically developed countries where prices for farm products are inflated by government subsidies and guarantees. Moreover, in economically developed countries there is a greater demand for higher-value livestock and horticultural products, compared to cheaper cereals and roots in the developing world.

The differences in output in relation to land area are influenced by: environmental conditions, management skills, farming systems, and **inputs** of **agrochemicals** and machinery. However, differences in output compared to the agricultural workforce are closely related to economic development and the percentage of the workforce in agriculture.

1.5 Types of agriculture

There are many ways by which we can recognise different types or classes of agriculture. Perhaps the most obvious is to use the dominant crop or enterprise, as in dairy farming or viticulture. There are, though, a great many other ways of classifying farming systems (Table 1.3). Needless to say, there is no one label which can describe all of the characteristics of any farming system. It may be convenient to do this, but, as we shall see in later chapters, each farming system has a range of characteristics. The most detailed classification schemes rely on **crop combinations** defined by statistical methods. You can find out more about this in Chapter 2.

Table 1.3 Characteristics of farming systems

	Hill sheep	Maize
A Dominant crops/enterprises		
arable		●
livestock	●	
mixed		
horticulture		
B Labour inputs		
high – intensive		
low – extensive	●	●
C Capital inputs		
high – intensive		●
low – extensive	●	
D Technology inputs		
high		●
intermediate	●	
low		
E Markets		
commercial	●	●
subsistence		
F Social organisation		
individual	●	●
collective		
co-operative		
tribal		
G Political organisation		
capitalist	●	●
redistributive/socialist		
H Settlement		
sedentary	●	●
nomadic		
shifting		
I Climate		
tropical	●	●
sub-tropical		
temperate		
J Moisture control		
irrigated		●
rain-fed/dry	●	
K Tenancy		
owner-occupier	●	●
tenant		
employee		

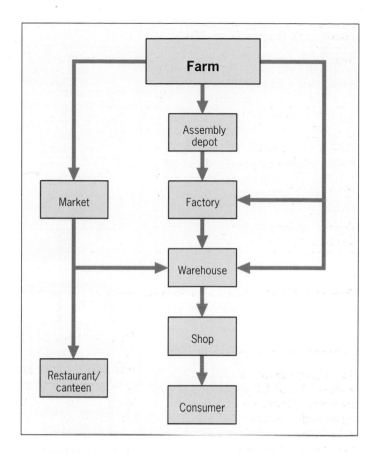

Figure 1.7 A food system (above right)

Figure 1.8 A lower level of economic development results in a shorter food chain: fruit market, Vietnam (below right)

1.6 What are food systems?

Food systems are formal and informal chains of people and institutions through which society gains its food supplies. We have already seen that farmers have a prominent role in food systems. They are the production side and therefore the starting point of the food chain. However, beyond the farm gate there are a number of **downstream functions** through which food must pass before reaching the consumer (Fig. 1.7). Generally, the higher the level of economic development, the longer such chains tend to be (Figs. 1.8 and 1.9 and Chapters 8 and 9). Some food shortages and famines may result not only through poor harvests but also by disruptions to the food chain (Chapter 10). Responses to food supply problems (e.g. food aid, technology, government policies) are the focus of Chapters 11 and 12.

?

6a Draw a simplified version of Figure 1.9, similar to Figure 1.7. (You will need to extend the chain downstream and refer to Figure 1.7 to add extra functions).
b What happens to the bulk of the corn crop? Which livestock depend most on corn for feed?

7 Try constructing your own food system diagram for either a hamburger or a packet of potato crisps.

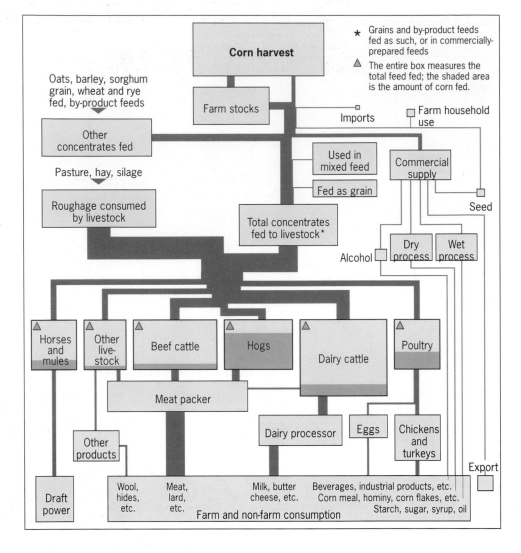

Figure 1.9 Corn (maize) food system, USA (*Source:* Kohls, 1967)

Summary

- Agriculture is part of a food system which includes farmers, wholesalers, food manufacturers, retailers and consumers.

- Agriculture is the control and use of plants and animals for the production of food, fibre and raw materials for industry.

- Agriculture has additional functions, including providing employment, food security, and recreation and leisure.

- Agriculture is fundamentally different to other economic activities like manufacturing and services.

- The importance of agriculture can be measured in terms of the number of agricultural workers, value of output, its use of land, and the issues to which it gives rise.

- At a global scale the spatial distribution of agricultural employment, output and land is highly uneven.

- Many different criteria are used to define types or classes of agriculture.

- Food systems are informal chains of people and institutions through which society acquires its food supplies.

- These food chains are generally much longer in economically developed countries than in economically developing countries.

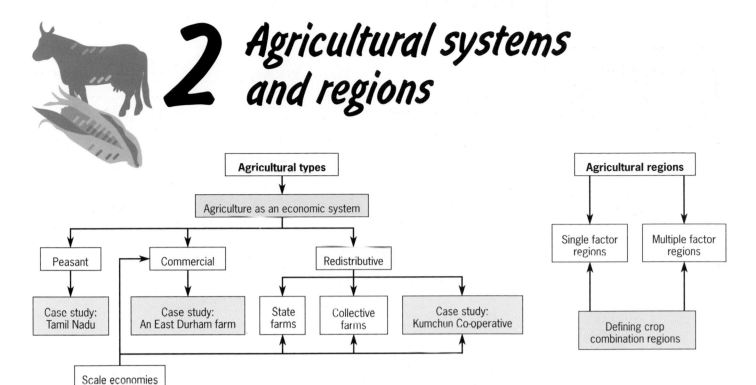

2 Agricultural systems and regions

2.1 Introduction

In this chapter we examine types of farming and their geographical distribution. We shall also see how geographers have tried to make sense of these distributions by defining **agricultural regions**.

2.2 Agricultural types

The simplest classification divides agriculture into three types: **peasant**, **commercial**, and **redistributive**. This classification is based on economic, social and political factors. Peasant agriculture is found in pre-industrial societies where the family, local community or tribe is the most significant unit. This type of agriculture was widespread in Europe before the nineteenth century and still dominates the developing world. Commercial agriculture operates within a **capitalist** framework. It is typical of industrial and post-industrial societies in the developed world. Redistributive agriculture is confined to countries like China, Cuba and North Korea which have **centrally-planned** or **command economies**.

Agriculture as an economic system

One way of thinking about agriculture is to view it as a **system** (Fig. 2.1). We can define a system as a set of parts and their characteristics, and the relationships between them. One advantage of systems is that they emphasise to us how things are connected.

We have chosen to study agriculture as an economic and a biological system. (We shall postpone our study of biological systems until the next chapter.) As an economic system a farm consists of an area of cultivated land and its farm buildings, and a series of **inputs** and **outputs**. In order to survive, the farmer must make a reasonable surplus or profit, and for this to happen s/he must ensure that the value of outputs (crops and livestock) exceeds the value of inputs (e.g. labour, machinery, seeds).

Agriculture as an
economic system

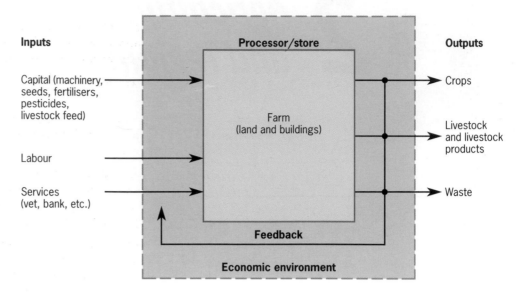

**Figure 2.1 Agriculture as an
economic system**

The farmland is the main **store** or **processor** in the system, and it has a
number of important features. These include the farm's size or area; its soils
and their fertility; its location in relation to markets, and so on.

The most important economic inputs into a farm system are labour and
capital. Small farms usually depend entirely on family labour, whereas large
farms will probably hire extra workers. Capital inputs include machinery
such as tractors, combine harvesters and grain driers, as well as on-going
expenditure on seeds, fertiliser, pesticides, livestock, diesel fuel and many
other items. The precise nature of these inputs will depend on levels of
technology in the society.

Economics and politics

Agricultural systems do not exist in isolation: they are part of a wider
economic and political environment. Thus commercial farming operates
within a capitalist structure where profit is the driving force and prices
depend on supply and demand. Modern commercial agriculture works within
a capitalist environment which is often worldwide.

On the other hand, redistributive agriculture is less concerned with profit.
Although it too must produce a surplus, it also aims to achieve equality and
social justice for farmers. Unlike commercial farming, it tends to operate at a
national scale with fewer trading links with the outside world.

Peasant agriculture operates mainly at a local scale. Indeed, where peasant
farming is geared to **subsistence**, most output does not even leave the farm.
However, in practice the vast majority of peasant farmers sell or exchange
some of their produce in local markets.

2.3 Peasant agriculture

A quarter of the world's population lives in peasant farming families. Peasant
farming has a number of features which make it quite different from
commercial farming:

- Farm holdings are usually small.

- Cultivation often occurs in difficult environmental conditions where there
is a high risk of crop failure or natural disaster.

- Most peasant families are poor, or very poor.

- Food production is partly for family needs (subsistence), and partly for the market. (Pure subsistence farming where nothing is sold off the farm, and no inputs such as seeds and fertilisers are bought, is rare today.)

- A peasant farmer's chief objective is to survive by satisfying short-term, basic needs.

- Peasant holdings are often very productive in terms of output per hectare. This is due to the intensive use of family labour, using simple tools, rather than machinery or chemicals.

- Although lacking formal education, peasants are often very knowledgeable about farming and the environmental conditions which affect their farm. The stereotype of the 'ignorant and backward peasant' is unfair: peasants are prepared to experiment provided that their survival is not threatened.

- Peasant farmers grow a wide range of crops. This is because crop diversity both ensures an adequate and varied diet, and helps to maintain soil fertility. In addition, a variety of crops helps to reduce the risk of crop failure.

Peasant farming in Tamil Nadu, India

Figure 2.2 India's southern state: Tamil Nadu

Figure 2.3 Subsistence agriculture in Tamil Nadu: threshing rice

By Indian standards the climate of Tamil Nadu (Fig. 2.2) favours agriculture. Rainfall averages 700 mm a year and comes from two monsoons: the north–east monsoon (October–December) and the south–west monsoon (June–September). Even so, the area has experienced devastating famines, and cyclones and droughts are a constant threat. There is, however, a high population density, though poverty is widespread, especially among landless labourers.

Irrigated rice (paddy) is the principal crop (Fig. 2.3). Water for **irrigation** is either taken from rivers or small reservoirs known as tanks. In Tamil Nadu farmers recognise three rice-growing seasons:

kuruvai: late June – late September
thaladi: late October – February
samba: August – January.

Kuruvai and *thaladi* can be grown in the same field, which allows double cropping each year. *Samba* takes longer to grow and is therefore a single crop. In addition to rice, farmers cultivate sugar cane, groundnuts, beans and fruit.

Mr Nagamuntha is a fairly typical poor peasant farmer: he is a *paraiyar*, one of India's lowest caste (class) groups, and married with three grown-up children. None of his children, however, has married and they remain on his farm. Although Mr Nagamuntha owns his farm, it is tiny – just 0.3 hectares – and even with irrigation and farming the double-cropped *kuruvai* and *thaladi*, it is too small to provide full-time employment for himself and his family. So, to make ends meet, Mr Nagamuntha sells

Tamil Nadu

some coconuts from his trees, some eggs and some chickens. He and his wife are also forced to work as farm labourers, pulling, transplanting and harvesting rice. The work is hard and their wages are poor.

In 1985 Mr Nagamuntha harvested 910 kg of *kuruvai*. Most of it was sold to meet various living expenditures. However, drought threatened his *thaladi* crop and reduced his final output.

Many peasant farmers in Tamil Nadu, including Mr Nagamuntha, grow new high-yielding varieties of rice, using chemical fertilisers and pesticides (see Section 11.4 for the development of high-yielding varieties). Unfortunately, although more productive, these new crops have disadvantages. For example, they require more water than the traditional varieties, and if water is scarce then fertilisers have only a limited effect.

Table 2.1 All-India crop yields (kg/ha/year)

	Traditional methods	Semi-modern methods
Rice	1500	3500
Wheat	1000	2500
Maize	1000	2500
Millet	700	2500

1a Calculate yields of rice per hectare for Mr Nagamuntha. Look at Table 2.1 and compare this yield with the overall averages for India.
b Draw a systems diagram (similar to Fig. 2.1) to describe Mr Nagamuntha's farm.
c Construct a table to show how Mr Nagamuntha's farm compares with the typical peasant farming system described in the theory section on page 14.

2 Study the information in Table 2.2 and complete the following exercises. (You should assume that two tonnes of grain per person per year are needed for survival on the peasant farm.)
a State two ways in which you could measure the intensity of cultivation (i.e. the ratio of inputs to outputs).
b Imagine a situation on a peasant farm, where the family size increased from five to six people. Suggest two possible responses which would enable food output to be increased.

2.4 Commercial agriculture

Commercial agriculture (Fig. 2.4) dominates farming in economically developed countries. Typically, it is large-scale, mechanised, and technologically advanced. The main goal of commercial farmers is to maximise profits. Thus crops and livestock are produced for cash and most of the output is sold off the farm. Unlike most subsistence farmers, commercial farmers buy inputs such as fertilisers, seeds, pesticides and animal feed from outside. Consequently, in commercial systems capital inputs are high compared to labour inputs. Because they are linked to the world economy through trade, commercial farmers are able to specialise more than peasant farmers. This brings certain advantages; the most important of these are **economies of scale** obtained by the bulk purchase of inputs, and cost savings from the use of machinery full-time.

Table 2.2 Comparison of a peasant and a commercial farm

	Peasant farm	Commercial farm
Farm size (ha)	2	200
Family size (persons)	5	5
Work days/year	1500	600
Cereals yield (tonnes per ha)	5	3
Output (tonnes per person)	2	120

Figure 2.4 Commercial agriculture in Queensland, Australia: pineapple plantation

A commercial farm in East Durham, England

This cereals and beef farm (Figs 2.5 and 2.6) is located on the Magnesian Limestone plateau of East Durham, approximately 125 metres above sea level. The physical conditions here are favourable for agriculture.

In total the farm comprises 64.5 ha. Of this, 36.5 ha are owned jointly by two farmers. The remainder, which is detached from the main farm, is rented.

Figure 2.5 A farm in East Durham, England

- Mean annual precipitation is nearly 700 mm
- Soils are light, sandy loams and are easily worked.
- Farmland quality is grade 3 (see table 3.9).

Not all the 64.5 ha are cultivated: Buildings and roads account for 6 ha, and there is a small area of mining subsidence which is poorly drained.

Figure 2.6 Farm layout, East Durham

Farm activities

The main enterprises on this farm (which is jointly owned in partnership by two farmers) are cropping and beef cattle (Fig. 2.7). However, cropping – primarily wheat and barley for animal feed – is the dominant enterprise. Other crops include winter oats, and oilseed rape, which is grown under contract to wholesalers. In addition, the farmers grow grass as pasture for grazing and for silage. (Silage is when the grass is cut green and sealed in polythene to exclude air. This prevents deterioration while sugars in the grass convert to lactic acid preserving the grass, thus enabling the farmer to store it for winter feed.)
Overall, there is a four-year sequence of cropping on each field, known as crop rotation: one year each of cereals and rape, followed by a two-year ley of sown grass.

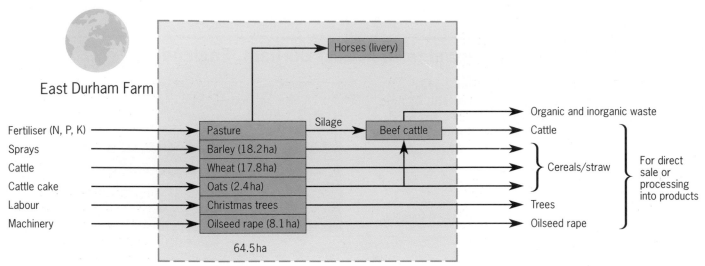

Figure 2.7 Farming system: East Durham farm

The farm is highly mechanised, and includes three tractors and a second-hand combine harvester. As the farmers cannot afford new machinery, they tend to buy any extra items from farm sales. Even so, the farm is unusual because it has no bank borrowings or overdrafts.

Farm profits

Because one of the farmers is beyond the normal age of retirement, they have decided not to grow labour-intensive crops, such as potatoes and turnips. In fact, the last full-time labourer left the farm in 1970. This contrasts with the 1950s when there were three full-time male and three full-time female workers on the farm. Today, the owners have difficulty in finding casual labour and so they hire contractors to cut hedges and wrap silage.

The farm used to have sheep and pigs, but its location close to urban areas led to problems such as animal theft, attacks by stray dogs and the smell of pig slurry when applied to the land. Instead, the farmers now keep 40 beef cattle each year for fattening. These are fed silage with a supplement of crushed barley and oats.

In recent years, low profits from the traditional farm activities have forced the farmers to diversify. They have since added three new enterprises:

1 They grow Christmas trees as part of a consortium of farmers who are aiming to undercut the price of imported trees. The trees take approximately seven years to mature.

2 Horses are grazed on the farm which offers DIY livery (stabling for horses which are looked after by their owners). This has become an important extra source of income.

3 The farmers supply snow-ploughing under contract for the local council.

Table 2.3 Traditional sources of income: East Durham farm, 1991

Crops	Area (ha)	Yield (t/ha)	Price (£/t)	Income
Wheat	17.8	7.3	109.9	?
Barley	18.2	6.2	108.8	?
Oats	2.4	5.5	106.1	?
Oilseed rape	8.1	1.4	343.9	?
Cattle	?	?	?	7523
Straw	?	?	?	2200

Table 2.4 Yearly balance sheet for traditional enterprises

	Costs	Income
Nitrogen fertilisers	3174	
Phosphorus/potassium fertilisers	2722	
Sprays	4261	
Cattle	3495	
Cattle cake	481	
Farm produced feed	500	
Veterinary bills	190	
Silage (harvesting)	1600	
Transport (including fuel)	60	
	16 483	16 483
Total farm gross margin:		

?

3 We can analyse the profitability of a farm by drawing a balance sheet of costs and income (Tables 2.3 and 2.4). If income exceeds costs, then the farmer makes a profit; if costs exceed income, then the farmer makes a loss.
a Using the information in Table 2.3, calculate the yearly income of the farm from traditional enterprises.
b Complete the balance sheet in Table 2.4 and calculate the yearly profit for the East Durham farm.
c What factors might have influenced the decision to diversify into horse livery and to grow Christmas trees?

4 You are a UN inspector for agriculture. Write a brief report aimed at a general readership explaining the similarities and differences between commercial and peasant farming systems in: farm size, labour intensity, capital intensity, range of crops grown, output per hectare and output per person.

2.5 Redistributive agriculture

In socialist or command economies, such as China, Cuba, and North Korea, agriculture is controlled by the state rather than individual farmers. This places decision-making in the hands of politicians and bureaucrats. As a consequence, agriculture reflects targets set by the state and not the preferences of individual farmers. There are three kinds of farm in redistributive systems: **state farms**, **collectives**, and **co-operatives**.

State farms

State farms are run like a state-controlled manufacturing industry, with a manager and hired workers: the farm workers are merely paid employees and have no say in the crops grown or cropping methods used. The main advantage of state farms is their size. They are very large (in the former USSR they averaged 19 100 ha) and this enables them to achieve scale economies comparable to large industries. Their main disadvantage is that they offer little incentive to workers to improve their output or efficiency.

Collective farms

Collective farms give workers greater interest and more say in their farm's success. Although the land remains state-owned, it is leased permanently to the farm workers who run the collective as a single unit, with labour organised into teams. The workers then share the profits and, in many cases, the farm workers are allowed to cultivate small private plots and sell their produce on the free market. However, as on state farms, production quotas are set by government.

Redistributive farming: the Kumchun Co-operative, North Korea

North Korea has a communist government. The country's development since the communist revolution in 1946 has focused on self-sufficiency and the use of the its own resources. Following the communist take-over, private farms were nationalised and today virtually all agricultural production comes from state or collective farms. Although the state gave a high priority to agriculture, production targets were still not met and this resulted in significant food imports during the 1980s.

The Kumchun Co-operative (Fig. 2.8) is situated near P'yongyang, the capital of North Korea. It employs 1833 people (organised into five work teams), and comprises 440 hectares of cultivated land. The main crops are rice, vegetables, fruit and maize (Fig. 2.9).

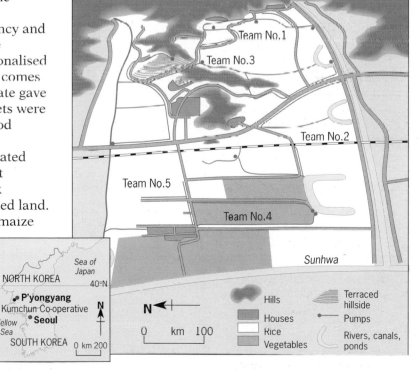

Figure 2.8 Kumchun Co-operative Farm

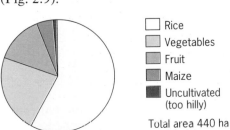

Figure 2.9 Kumchun Co-operative: pattern of cropping

Kumchun Co-operative

Although cultivation is labour-intensive, the farm is well provided with machinery and has 32 tractors, 15 trucks, and 6 harvesters. Individual farmers are also allowed to cultivate small kitchen gardens (100m² per household) for their own vegetables, and each household may keep up to ten ducks or chickens and two pigs. Day-to-day decisions are made by a chairperson who is elected annually by the workers.

The fertile soils and reliable irrigation give good yields, which are well above world averages (Table 2.5).

Workers are paid in cash and their wages are in proportion to their work effort and skill. The wages are decided once a year on Labour Day when the work points earned are totted up. For example, a day's rice transplanting is worth 1.6 points, while a day's weeding earns 1.4 points. If we assume that everyone's work points total 100 000 and the co-operative produces 1000 tonnes of rice, then each point is worth 10 kg of rice (which is a fixed price).

In 1991 the net income per head amounted to just £360. However, this figure takes no account of work done in exchange for goods or services. Nor does it include the free medical care, child care and housing received by the workers. Food and fuel are also heavily subsidised. For example, in 1991 the state was selling rice at just one-eighth of its real cost. None the less, failure to achieve production targets means that rice and other essential commodities are still rationed.

Table 2.5 Rice and maize yields: Kumchun and world averages (Yield/ha:tonnes)

	Kumchun	World
Rice	9.62	3.56
Maize	7.50	3.68

Figure 2.10 Kumchun Co-operative: workers' accommodation

5 Study Figures 2.8 and 2.10.
a Draw a systems diagram to describe the Kumchun Co-operative.
b Research the topic of **intensive farming** (using textbooks, articles, databases and any other sources). Focus on the following areas: water management, soils, wildlife, and landscape.

6 Essay: Compare and contrast the economic, social and technological advantages of two types of farming.

2.6 Agricultural regions

Having described the types of agriculture in Sections 2.2–2.5, we can now turn our attention to their geographical distribution. If you look at Figure 2.11, you will see that the distribution is far from random. It appears that similar farming activities concentrate in specific areas.

Geographers find it useful to think of these concentrations as agricultural regions. Thus, when we divide up the earth's surface into regions, we call the process **regionalisation**. This process is like classification in science. Zoologists, for example, group animals with similar characteristics into classes such as insects, fish, birds, mammals and so on. In the same way, geographers recognise areas with similar crops and farming methods as agricultural regions.

Regionalisation is an effective way of simplifying complex spatial patterns. However, we must recognise that it involves generalisation and some loss of detail. Moreover, you lose more detail as the scale becomes larger.

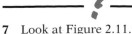

7 Look at Figure 2.11.
a Make a list of the criteria used to define types of agriculture across the globe (e.g. crop types).
b Why is it so difficult to define agricultural regions using a single factor such as land use?

Figure 2.11 World farming regions, based on Perpillou's classification of agricultural types

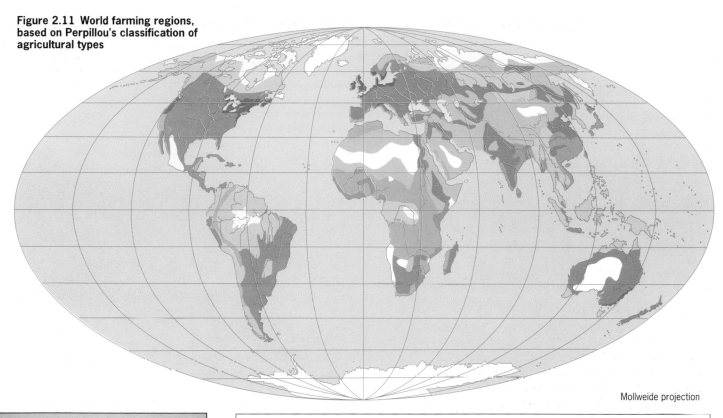

Mollweide projection

☐ Primitive collectors (gathering, hunting, fishing)	☐ Intensive mechanised arable farming, linked with intensive stock-raising and other livestock enterprises (e.g. dairying)
☐ Pastoral nomads	☐ Intensive arable farming based on human labour and rice-growing
☐ Forest dwellers (hunting, trapping or shifting cultivation)	☐ Mediterranean agriculture and tree-cropping
☐ Forest dwellers (exploitation of timber)	☐ Plantation agriculture in tropical areas
☐ Primitive extensive agriculture, not linked with stock-raising	☐ Extensive stock-raising and ranching

Figure 2.12 Harvesting in the US corn belt

Table 2.6 UK farm enterprise classes defined by the MAFF

Dairy cattle
Beef cattle and sheep
Cropping
Pigs and poultry
Horticulture
Mixed

2.7 Agricultural regions in England and Wales

Single-factor regions

The simplest regionalisation is based on a single factor, usually land use. A classic example is the US corn belt (Fig. 2.12) where a single crop is used to define the region.

Figure 2.13 shows the agricultural geography of England and Wales as a set of single-factor regions. These regions are defined by the Ministry of Agriculture Fisheries and Food (MAFF) according to the dominant farming enterprise in each county. The dominant enterprise is the farming activity which uses the largest amount of labour input. This is measured in standard work-days per year.

The MAFF groups farms into one of five types where 50 per cent or more of standard work-days per year are devoted to a particular enterprise. Where no enterprise accounts for 50 per cent or more of standard work-days, it is allocated to a sixth category called 'mixed'. The six enterprises are listed in Table 2.6.

Regionalisation: weaknesses

Given the level of generalisation, we should not be surprised by the broad

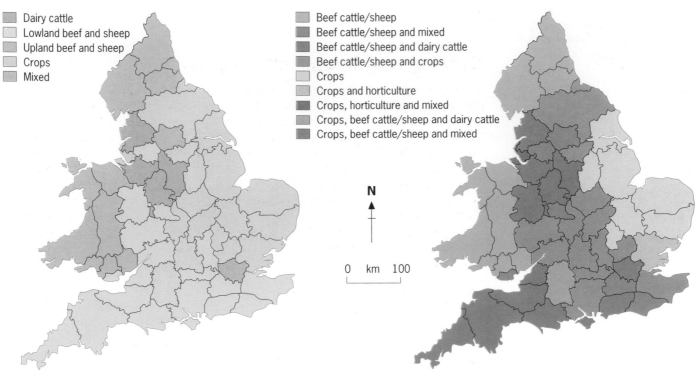

Figure 2.13 Agricultural regions: dominant enterprises in England and Wales

Figure 2.14 Agricultural regions: enterprises accounting for 50 per cent or more of holdings in England and Wales

8 Describe in detail the spatial pattern of farming in England and Wales shown in Figure 2.13.

9 Study Figure 2.14 and answer the following questions:
a How many different crop combinations are needed to cover all 52 counties in England and Wales?
b Six crop combinations account for 47 of the counties. Name these six crop combinations.
c State one advantage and one disadvantage of the regionalisation of Figure 2.14 compared to Figure 2.13.

spatial patterns shown in Figure 2.13 – with most counties forming parts of larger regions. Even so, we must recognise the weaknesses of such a regionalisation. The main ones are:

1 The dominant enterprise in some counties makes up less than one-third of all farm holdings. For example, Kent has a particularly varied agriculture. Although lowland beef cattle and sheep is the leading enterprise here, it is dominant on just 27 per cent of holdings. Indeed, only eight counties have a single enterprise which is dominant on 50 per cent or more of all farm holdings.

2 Counties are too uneven to be useful for generalisation. They vary widely in area, and the largest have a broad range of internal variations in relief, climate and soils. For instance, the largest county, North Yorkshire, could be divided into smaller sub-regions such as the Dales, North York Moors and Wolds (beef cattle and sheep), and the Vale of York and Vale of Pickering (cropping).

3 Because the MAFF statistics are only available for counties, Figures 2.13, 2.14 and 2.15 suggest abrupt changes in farming types at county boundaries. This impression is wholly artificial.

Multiple-factor regions
We can overcome some of the weaknesses of regionalisation seen in Figure 2.13 if we use a multiple-factor approach. After all, the main feature of agricultural distribution is that crops are usually grown in combinations. Thus, lowland beef and sheep enterprises in the UK are often combined with dairying in the west and arable farming in the east. Figure 2.14 takes regionalisation of agriculture a step further by defining **crop combination regions** which account for 50 per cent or more of all farm holdings.

The most detailed definition of crop combination regions relies on a statistical technique and the calculation of variance.

Defining crop combination regions

We can define crop combination regions by calculating the variance of the percentage of actual holdings of each enterprise from the theoretical percentage. Thus, if the crop combination in a region (which had six enterprises) were best described by **monoculture** (i.e. one crop) then the theoretical distribution of holdings attached to each of the six enterprises would be 100, 0, 0, 0, 0, 0. A two-crop combination system would be 50, 50, 0, 0, 0, 0, and a three-crop combination 33.3, 33.3, 33.3, 0, 0, 0, and so on. If we calculate the variance of the actual distribution compared to the theoretical, then the *lowest* variance gives us the combination of best fit. An example is given in Table 2.7 for Lothian in eastern Scotland, which shows that the pattern of farming in the county is most accurately described by the two-crop combination of mixed and cropping (lowest variance).

Table 2.7 Actual and theoretical distributions of farm enterprises in Lothian

	Actual distribution	Theoretical distribution						Variance
	(%)	1	2	3	4	5	6	
Mixed	43.1	100	0	0	0	0	0	779.7
Cropping	35.1	50	50	0	0	0		79.6
Beef cattle/sheep	13.2	33.3	33.3	33.3	0	0	0	169.4
Dairy cattle	5.2	25	25	25	25	0	0	161.4
Pigs/poultry	2.6	20	20	20	20	20	0	221.7
Horticulture	0.7	16.7	16.7	16.7	16.7	16.7	16.7	272.5

Step 1: List the theoretical (T) and actual (A) distributions in paired values.

Step 2: Subtract the actual distribution from the theoretical for each pair of values.

Step 3: Square the differences (D) between theoretical and actual distributions.

Step 4: Sum the squared differences (D^2) and divide the sum by the number of paired values to find the variance.

?

10a Find the combination of crops which best describes farming in Fife (eastern Scotland) and Dumfries and Galloway (South-West Scotland) by calculating the variances for the enterprises in Table 2.9.
b Suggest reasons for the differences in farming between the two counties.

Table 2.8 Example: calculation of variance for two-crop combination

T	A	D	D^2
50	43.1	6.9	47.61
50	35.1	14.9	22.01
0	13.2	−13.2	174.24
0	5.2	−5.2	27.04
0	2.6	−2.6	6.76
0	0.7	−0.7	0.49
			478.15

Variance $= \dfrac{478.15}{6} = 79.69$

Table 2.9 Farming enterprises in Fife and Dumfries and Galloway (per cent holdings in each enterprise)

	Fife	Dumfries and Galloway
Dairy cattle	6.18	25.1
Beef cattle/sheep	13.9	40.0
Cropping	43.5	3.5
Pigs/poultry	2.3	0.3
Horticulture	0.3	0.1
Mixed	33.1	31.0

Figure 2.15 shows the classification of each county in England and Wales using the variance method we described above. Its most striking feature is the great variation in the number of enterprise combinations. Thus while we can best describe agriculture in Cambridgeshire by a single enterprise, the West Midlands is so varied that a six-enterprise combination is needed. However, in the majority of counties, agriculture fits a three- or four-enterprise combination.

Figure 2.17 Dairy farming: Clwyd, north Wales

Figure 2.16 Upland sheep farming: Black Mountains, south Wales

Figure 2.15 Agricultural regions: crop combinations in England and Wales

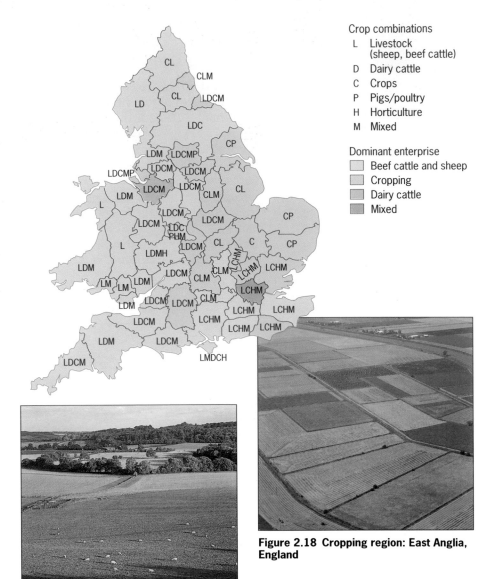

Crop combinations
L Livestock (sheep, beef cattle)
D Dairy cattle
C Crops
P Pigs/poultry
H Horticulture
M Mixed

Dominant enterprise
Beef cattle and sheep
Cropping
Dairy cattle
Mixed

Figure 2.18 Cropping region: East Anglia, England

Figure 2.19 Mixed farming: Buckinghamshire, England

?

11 With reference to Figures 2.15–2.19, describe the main characteristics of the following agricultural regions: South-East; eastern England; North-West and South-West; Wales and the North.

12 Which are the most specialised agricultural regions in Figure 2.13? Suggest possible reasons for this specialisation.

13 Using Figures 2.15–2.19, comment on the distribution of five- and six-crop combination types. Suggest possible explanations.

14 Compare the different regional patterns of farming in England and Wales in Figures 2.13, 2.14 and 2.15. Which would you say is the most effective? Keep in mind the need for accuracy and detail, as well as broad spatial patterns.

Summary

- At a global scale we recognise three broad systems of agriculture: peasant systems; commercial systems; and redistributive systems.

- Peasant agriculture is confined to economically developing countries and is centred on the family and tribe. It relies on simple technology and limited exchange.

- Commercial farming is often large-scale and technologically advanced. It operates within a capitalist system and aims to maximise profits.

- Redistributive agriculture is found in socialist countries. Unlike other farming systems, it is subject to strong centralised control. It aims as much at social justice as efficient food production.

- At a national scale, with uniform political economy and levels of development, differences in agricultural systems reflect a farmer's choice of enterprise.

- From a geographical perspective, these farm enterprises often show a marked spatial patterning and give rise to distinctive agricultural regions.

3 Managing biological systems

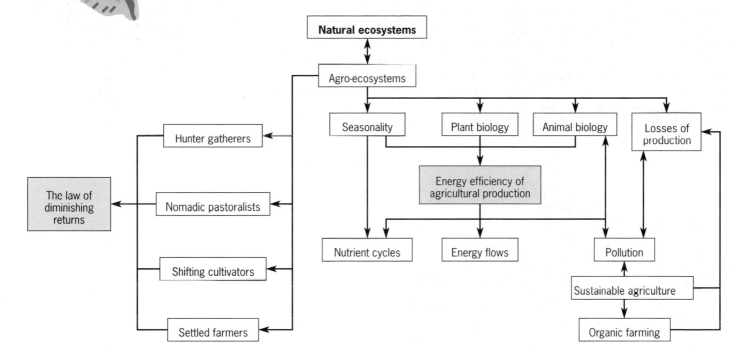

3.1 Agriculture as an ecosystem

In Chapter 2 we studied agriculture as an economic and political **system**. Here we examine agriculture as a biological system or **ecosystem**. In an ecosystem there is an interdependence, and an interaction between, plants and animals and their physical, chemical and biological environment. The flows of energy and nutrients link the living and non living parts of the ecosystem. We may study ecosystems at different scales, depending on our purpose: from an individual farm (Fig. 3.1) to major world **biomes** such as the tropical rainforests or temperate grasslands.

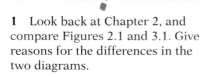

1 Look back at Chapter 2, and compare Figures 2.1 and 3.1. Give reasons for the differences in the two diagrams.

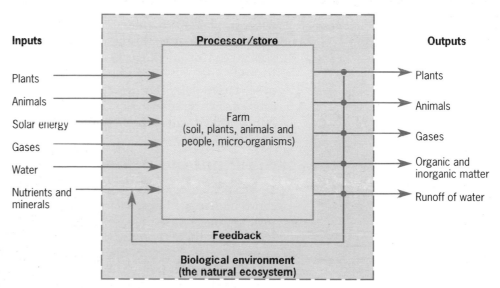

Figure 3.1 Agriculture as a biological system

Figure 3.2 Tuareg nomadic pastoralists, Timbuktu, Mali

Figure 3.3 Settled agriculture, Burkina Faso

?

2 Study Figures 3.2 and 3.3.
a For each photograph, list as many examples as you can find of modifications to the natural ecosystem.
b Compare and explain the differences between the two types of agriculture, using systems terminology (inputs, outputs, **processor/store**, feedback).

3a A variety of crop types are mentioned in *Agriculture and Food*. To assess their usefulness and value to humans, visit your school or local library and research the reproduction details of the following: rice, wheat, potatoes, sugar cane, cassava, cocoa, maize, coffee and coconuts.
b Group each crop into one of the four classes in Table 3.1.

Farmers exploit ecosystems and their activities may or may not be **sustainable**. **Nomadic pastoralists**, for example (Fig. 3.2), have traditionally managed natural ecosystems so that the **inputs** balance the **outputs**. In agricultural terms, provided the pastoralists keep the densities of livestock low, their nomadic way of life can be sustained indefinitely.

Settled agriculture is different (Fig. 3.3). It transforms natural ecosystems by deliberately modifying plants, animals and the environment for food production. We refer to these human-made ecosystems as **agro-ecosystems**.

A major difference between agriculture and other industries is agriculture's focus on living plants and animals. Such items are more short-lived and harder to manage than machines in factories, and cause difficulties for farmers. We can summarise these problems as: seasonality; the nature of plant and animal biology; and losses of production.

3.2 Seasonality

Everywhere outside the equatorial regions, seasonality is a fact of farming life. As we move polewards the seasons – warm and cold, wet and dry (see Chapter 4) – become more pronounced. Because most seasons are predictable, farmers can prepare in advance for the arrival of spring or the start of the rainy season. This allows them to take advantage of the growth surge of plants and animals which occurs at this time.

Seasonality often means that periods of glut are followed by periods of shortage. This is evident in the seasonality of milk production: our demand for milk and milk products is relatively constant throughout the year. But in grass-based dairy farming there is a sudden peak in milk output as the spring grass grows. Farmers try to even out such peaks by supplementing their animals' feed with silage and concentrates at other times of the year.

3.3 Plant biology

Domesticated plant species, or cultivars, have an extraordinary range of reproductive methods. These fall into two groups: sexual, involving the production of seed; and asexual (or vegetative) through stems, roots and leaves. It is these reproductive structures which humans can use, and which consequently determine the value of the plant. A further distinction is between **annuals** (which are sown and harvested within one year), and **perennials** (which, once planted, yield annually for several years).

Table 3.1 Crop types and reproduction

	Sexual	Asexual
Annuals		
Perennials		

Energy efficiency of agricultural production

We can measure the efficiency of natural ecosystems by comparing inputs of solar energy with the amounts converted to plant tissue in **photosynthesis**. Most natural ecosystems are highly inefficient: plants only trap between one and five per cent of insolation (incoming solar radiation). Despite its inefficiency, photosynthesis is vital because plants are the basic producers in all ecosystems. They convert solar energy to chemical energy, and it is in this form that energy becomes available for use by animals. Animals, such as caterpillars and deer, which eat plants directly, are **herbivores**. Meat-eating animals, or **carnivores**, consume plants indirectly by preying on herbivores. **Omnivores**, e.g. pigs, or humans, eat both plants and meat.

Figure 3.4 The food chain: energy flow within a general ecosystem (*After:* Chapman, 1993)

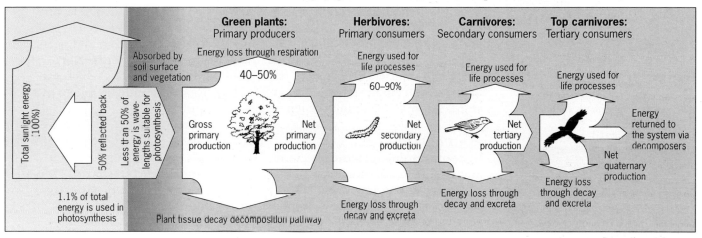

As animals consume plants and each other, energy is transferred along a **food chain** (Fig. 3.4). However, at each level in the food chain, most of this energy is transferred out of the system. Thus the total amount of energy in the bodies of rabbits is only a tiny fraction of that in grasses and herbs. Even so, rabbits comprise a greater energy store than stoats; and eagles, as the top carnivore, represent the smallest energy store in this food chain. It follows that the longer the food chain, the greater are the energy losses. As we shall see later, this has important implications for agriculture and food supplies.

This energy efficiency can be expressed as a percentage:

$$\text{Efficiency (\%)} = \frac{\text{useful output}}{\text{input}} \times 100$$

In agro-ecosystems we assess the efficiency of livestock in two ways: by their biological capacity to convert feed into energy; and by their ability to convert protein in their feed into protein in meat or milk (Table 3.2).

4 Look at Figure 3.4.
a How is energy transferred along the food chain?
b Describe the changes at each level in the food chain, and explain what effect this has on the top carnivores.

5 Which domesticated animal in Table 3.2 is:
a the least efficient,
b the most efficient converter of energy and protein?

6 In a peasant farming system such as Tamil Nadu in India (see Section 2.3), which animal would you keep? Briefly explain your choice.

7 Why would it be more efficient if crops replaced animals as the main food source for people?

Table 3.2 Efficiency of feed conversion in farm animals (*Source*: Spedding et al., 1981)

	Carcass output in kg / Feed intake in kg dry matter x 100	Total energy in carcass / Gross energy in feed x 100	Muscle protein gain / Protein intake x 100
Cattle	8.6	5.2–7.8	7.7
Sheep	13.2–17.6	11.0–14.6	13.3
Pigs	47.9	35.0	15.4
Hens	36.0	16.0	16.6

Figure 3.5 Keeping sheep close to farm buildings for protection against poor weather conditions, Pennines, West Yorkshire

Figure 3.6 Intensive turkey rearing unit, Edinburgh

?

8 Study Table 3.3 and complete the following exercises:
a How many offspring would you have in five years from one:
• cow, • sheep, • pig?
b Which of these animals might a farmer choose to rear, and why?
c Suggest what other factors might influence a farmer's decision whether or not to rear pigs (look back at Table 3.2.)

3.4 Animal biology

As with crops (see Chapter 4), there are optimal conditions for rearing domesticated animals. These conditions give the most efficient conversion of feed into products like eggs, meat, milk and wool. The domestic hen, for example, produces most egg weight at 13°C.

Farmers have greater control over the environment for their animals than for field crops. During the winter, and in spells of bad weather, livestock are kept in barns and outbuildings (Fig. 3.5). However, with modern factory farming, poultry, pigs and cattle can be kept indoors throughout the year (Fig. 3.6). This allows **intensive production** through the use of controlled temperatures, humidity, light, feeding and veterinary care. The nearest equivalent of this for plants is greenhouses.

Livestock farmers are faced with two important decisions. First they must decide which animals are best suited to the feed they have available; and second they must judge what economic return the feed will give in usable animal product. For example, ruminants, such as cattle, sheep and goats, can be reared on natural grasslands like the Russian steppes and the North American prairies. This is because of their complex stomachs which can digest fibrous materials such as grass. These foods are indigestible to more simple-stomached animals such as pigs and poultry.

Animal feed

In **commercial agriculture**, animal feeding has become increasingly artificial. The use of concentrated and formulated feeds gives farmers greater control over animals' diets, but at some cost: concentrates are expensive. They represent 30 per cent of total costs for intensive beef production: 50 per cent for milk and sheep: and 80 per cent for pigs. Obviously, therefore, livestock farmers must look closely at the efficiency with which each animal converts feed into meat or milk (Table 3.2).

Animal reproduction

When choosing livestock, farmers will also take account of the biology of animal reproduction. Characteristics such as breeding frequency, offspring per litter, gestation period and life-span will all influence the farmer's decision (Table 3.3).

Table 3.3 Reproductive characteristics of farm animals (Source: Spedding et al., 1981)

Animal	Breeding frequency (times/year or eggs/year)	Offspring (per litter)	Gestation (days)	Lifespan (years)
Cow	0.9	1	280	8–14
Sheep	1.0–1.5	1–3	147	5–8
Goat	1	1–3	150	6–10
Pig	1.0–2.2	8–12	114	2–3
Hen	180	–	21	2

3.5 Losses of production

Anyone with experience of vegetable gardening will know that the potential yield of a piece of ground is rarely achieved. Losses occur because some parts of the crops cannot be eaten. Similar limitations apply to agriculture, only on a much larger scale. Take the example of wheat: about 40 per cent of its dry weight is stalk and leaves; in brussels sprouts this rises to 70 per cent. No factory owner would accept such wastage in a manufacturing process.

Figure 3.7 African tsetse fly

?

---

9 Essay: Examine the principle of energy efficiency in agro-ecosystems with reference to peasant and commercial farming systems.

We can measure production losses in livestock by taking the ratio of high-value lean meat to low-value bone and gristle. This can be expressed as an equation:

$$\text{Production loss (\%)} = \frac{\text{h-v lean meat}}{\text{l-v bone and gristle}} \times 100$$

For cattle meat, the ratio is about 40 per cent of the carcass weight. The ratio for sheep is around 50 per cent, and for pigs and poultry 30 per cent. Clearly, there are economic rewards for any plant or animal breeder who can improve these ratios. This has led to the use of hormones and other steroids to increase weight in livestock.

The size of yield also depends on the health of crops and animals. It is estimated that 40 per cent of the world's wheat, and 50 per cent of the world's rice crop, are lost to pests and diseases. Animal diseases may be acute, such as epidemics of rinderpest, foot-and-mouth and fowlpest, or they may be much slower-acting, like bovine tuberculosis and 'mad cow disease' (BSE). In commercial economies, intensively-reared animals are often fed antibiotics as protection against some infections.

Animal parasites often weaken the livestock rather than kill them outright. For instance, gut parasites divert much-needed nutrition from the host. But while it is possible to control most animal parasites with suitable chemicals, it is difficult to eliminate disease-bearing insects like the tsetse fly (Fig. 3.7). This insect is responsible for nagana in cattle, and has made much of tropical Africa useless for cattle farming.

3.6 Natural ecosystems and agro-ecosystems

Natural ecosystems support three types of human economy: **hunter-gatherers**, nomadic pastoralists (Fig. 3.2), and **shifting cultivators** (Fig. 3.8).

Hunter-gatherers
Hunter-gatherers exploit natural ecosystems by hunting wild animals and collecting edible fruits, seeds, roots, tubers and leaves. Hunter-gathering is only sustainable where population densities are low.

Nomadic pastoralists
Nomadic pastoralists keep livestock which graze on natural grasslands. They must take care not to overstock fragile pastures and cause any deterioration in grazing resources. Such a delicate balance can only be achieved by a nomadic existence involving seasonal movement in search of pasture.

Shifting cultivators
Shifting cultivators live on small temporary plots cleared by burning areas of tropical rainforest or bush savanna vegetation. Their cultivation is based on modifying (rather than permanently transforming) the natural ecosystem. Each plot is cultivated for a year or two before being abandoned and reverting to forest. Like hunter-gathering and nomadic pastoralism, shifting cultivation is sustainable so long as population densities remain low. This allows the cleared plots to remain uncultivated for long enough to recover their fertility.

Settled agriculture
In contrast to these simple food production systems, settled agriculture transforms natural ecosystems. These agro-ecosystems differ from natural ecosystems in a number of ways (Table 3.4).

Figure 3.8 Yanomani Indians cultivating part of the Amazon rainforest

Table 3.4 Differences between agro-ecosystems and natural ecosystems (*After*: Tivy, 1990)

Agro-ecosystems	Natural ecosystems
Little diversity of plants and animals. Crops are often grown as **monocultures**.	Wide diversity of plant and animals species.
Simple structure with short food chains which focus on people. In livestock farming, people occupy the top carnivore niche. In arable farming they occupy the herbivore niche.	Complex structure with several levels in food chains: plants; herbivores; carnivores and top carnivores.
A large proportion of the biomass (mass of all living organisms in the ecosystem) comprises animals, especially large ruminants.	Only a tiny fraction of the biomass comprises animals.
Only a small proportion of energy finds its way into dead and decaying matter in the soil.	All dead organic matter is returned to the soil, decomposed and nutrients recycled.
Nutrient cycling is speeded up and is usually maintained by inputs of organic or inorganic fertiliser.	Nutrient cycle is slower. There are few inputs of organic matter from outside.
More 'open' than natural ecosystems, they exchange energy and material with the outside world. Thus animal feed, fuel and fertiliser may be imported: and waste products (e.g. polluting chemicals) may be 'exported' into other agro-, or natural, ecosystems.	Only limited movement of energy and materials across the ecosystem boundaries. Natural ecosystems are not subsidised by other ecosystems.

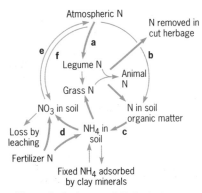

a Nitrogen fixation by symbiotic Rhizobium bacteria (root nodules)
b Nitrogen fixation by free-living bacteria
c Ammonification
d Nitrification
e Denitrification
f Additions of combined nitrogen from the atmosphere

Figure 3.9 The nitrogen cycle (*Source*: ICI)

10 You are a farm adviser who has to explain nutrient demand to a young farmer. Using the information in Figure 3.10:
a Describe the pattern of demand by crops for nitrogen, phosphorus and potassium.
b Compare the nutrient demands of arable and livestock farming. Briefly explain the differences.
c Suggest possible reasons why potatoes and maize have greater nutrient demands than wheat and rice.
d On the evidence of nutrient requirements in Figure 3.10, which crops would you say are least suited to peasant agriculture? Why?

3.7 Nutrient cycles

Plants need mineral nutrients which contain the chemical elements nitrogen (N), phosphorus (P) and potassium (K). Plants also need other minerals such as copper (Cu) and calcium (Ca), but only in tiny quantities. Most nutrient elements initially come from rocks and are released into the soil by weathering. However, nitrogen (N) is an exception.

Nitrogen
Plants and animals need nitrogen for growth and to repair their cells. Nitrogen occurs in air (it comprises about 78 per cent of air by volume), but plants and animals cannot use it directly. Plants can only absorb nitrogen when it combines with other substances to make compounds such as nitrates or ammonia. This combining and release of nitrogen is part of a cycle (Fig. 3.9). Nitrogen compounds which are not taken up by plants are easily washed out of the soil and pollute water courses.

Nutrient requirements
Overall, crops vary in their nutrient requirements (Fig. 3.12). Farmers therefore have the task of matching the demands of crops with the nutrient capability of their soils. However, this constraint is not as severe as it sounds: nutrients can always be added to soils as fertiliser.

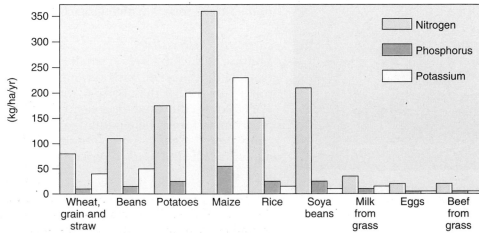

Figure 3.10 Nutrients removed in harvested crops and animal products (*After*: Spedding et al., 1981)

Nutrient cycling in natural and agro-ecosystems

There are many differences between nutrient cycling in natural and agro-ecosystems. One of the most important is that in natural systems, nutrients are re-cycled by decomposition of dead **organic** matter (Fig. 3.11): in agro-ecosystems most organic matter (and therefore most nutrients) is removed in harvested crops (Fig. 3.12). This means that farmers must subsidise agro-ecosystems with nutrient inputs – either as **inorganic** fertiliser or by purposeful re-cycling on the farm.

Farmers wishing to re-cycle nutrients have several alternatives. They may grow **legumes**, such as beans or clover, which have bacteria in their roots that can fix nitrogen from the atmosphere. Alternatively, they may plough in 'green manure' crops, or spread animal manure on to fields. In the UK, some sewage sludge is permitted on fields, although EC directives have limited this, as sewage can lead to toxic waste (mainly heavy metals) reaching the food supply. Similarly, in China, human sewage (or 'night soil') is used to fertilise vegetable fields around cities.

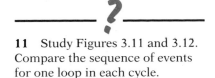

11 Study Figures 3.11 and 3.12. Compare the sequence of events for one loop in each cycle.

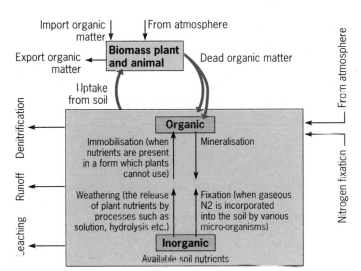

Figure 3.11 Nutrient cycling in a natural ecosystem (*After:* Tivy, 1990)

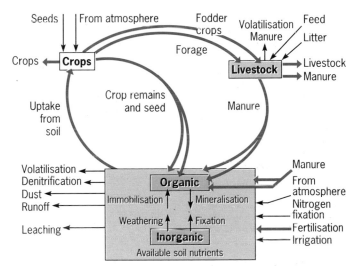

Figure 3.12 Nutrient cycling in an agro-ecosystem (*After:* Tivy, 1990)

3.8 Energy flows

In addition to nutrients, there is a continual throughput (flow) of energy in agro-ecosystems. The sun is the ultimate source of these energy inputs (Fig. 3.4). Plants capture solar energy directly through photosynthesis and store it as chemical energy. (This energy store is also used indirectly in the form of fuels, such as coal, oil or wood.)

Energy conversion

All plants require light energy for photosynthesis. This is the process by which the chlorophyll in green leaves produces carbohydrates from carbon dioxide and water. Less than three per cent of solar radiation falling on crops is converted into plant tissue; and this proportion is reduced even further before crops become human food.

Overall, energy conversion in agriculture for human food is inefficient. This becomes clear if we compare the food energy available from a potato crop with that from grass-based beef cattle. On average in the UK, only 0.25 per cent of the solar energy that falls on a potato field ends up as food for human consumption. But beef production is even less efficient: its conversion rate is a mere 0.02 per cent.

Table 3.5 Number of people whose annual energy and protein needs could be met from the output of 1 hectare (*After*: Spedding, 1981, 1988)

	Protein	Energy
Cabbage	34	23
Field beans	26	9
Peas	24	9
Sugar beet	17	33
Maize	16	17
Potato	22	22
Wheat	20	15
Barley	15	13
Rice	16	19
Milk	5	2
Chicken	6	2
Eggs	3	1
Mutton	3	2
Beef	3	1
Bacon	4	3

There is little doubt that, if we measure agricultural output as useful energy or protein per hectare, then arable farming is more efficient than livestock farming. Figures like those in Table 3.5 help to explain why some diets, which are dominated by cereals and have little or no animal protein, are perfectly adequate.

Livestock farming

Livestock farming is so energy inefficient that we might wonder why it is so popular. There are four possible answers. First, energy efficiency is of little interest to farmers as long as fuel for farm machinery and fertilisers are plentiful and cheap. Second, there are many other ways of measuring agricultural efficiency, such as returns on capital invested or labour inputs. Third, little cropping is possible on certain lands due to altitude and relief. For example, approximately 50 per cent of UK land is in Less Favoured Areas (LFAs) so livestock farming predominates. And, fourth, farmers might simply say that livestock products are more profitable. This is because many people in the developed world can afford to buy meat, milk, butter and cheese. Also, with rising incomes, people opt to consume more animal rather than plant products such as bread and potatoes (see Sections 10.3–10.4).

Indirect energy inputs

Our analysis of energy inputs is far from complete. Commercial agriculture indirectly imports huge amounts of energy into food production (Fig. 3.13). The most obvious is the use of hydrocarbon fuels to drive machinery. However, there are other hidden areas of consumption. Fertilisers, biocides and other **agrochemicals**, for instance, are manufactured by complex energy-hungry processes. The same is true of agricultural machinery.

After the products leave the farm there are further chemical energy inputs: transporting goods to market requires fuel; so does the factory processing of food and packaging. And even more energy is spent on distribution and retailing before the consumer makes a purchase (Fig. 3.14).

Figure 3.13: Direct and indirect energy inputs for breadmaking

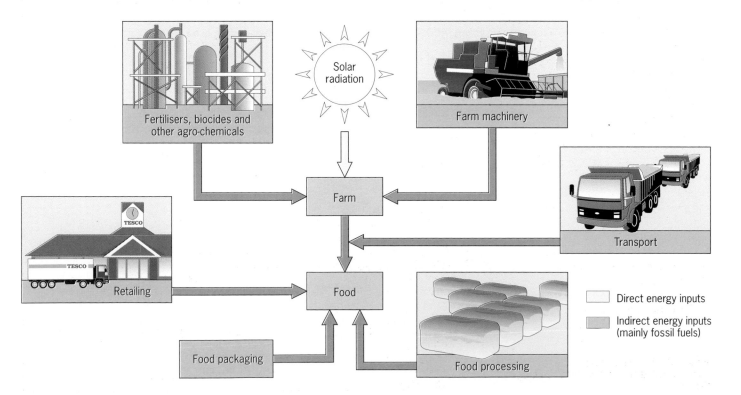

Figure 3.14: Energy requirements (per cent): 1 kg white loaf (*Source:* Leach, 1976)

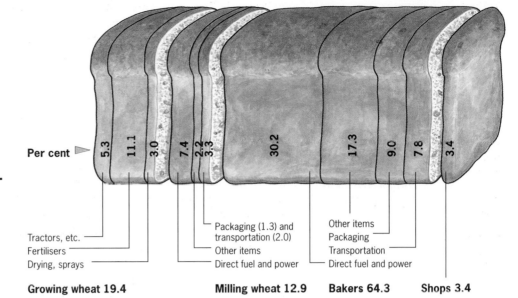

Per cent ▷ 5.3 | 11.1 | 3.0 | 7.4 | 2.2 | 3.3 | 30.2 | 17.3 | 9.0 | 7.8 | 3.4

Tractors, etc.
Fertilisers
Drying, sprays

Packaging (1.3) and transportation (2.0)
Other items
Direct fuel and power

Other items
Packaging
Transportation
Direct fuel and power

Growing wheat 19.4   **Milling wheat 12.9**   **Bakers 64.3**   **Shops 3.4**

?

12 Study the budgets of the five farms in Figure 3.15.
a Calculate the energy input/output ratios for the five enterprises.
b Which type of farming has:
• the most favourable, • the least favourable ratio? Suggest reasons for this.
c Describe the relationship between energy inputs and outputs for the four livestock enterprises.
d Explain why the energy ratio for the pigs and poultry enterprises is so much higher than the ratios for the other livestock and cereal enterprises.

13 We have seen that crop cultivation produces far greater amounts of protein and energy per unit area than livestock farming. We also know that a sizeable proportion of the world's population is inadequately fed (see Chapter 10). Consider the following statement:
'Wherever possible, we should restrict livestock farming to areas which are too dry or too cold for crops in order to use the remaining land for arable farming and thus maximise global food production.'
Prepare a reply to this statement from:
a a livestock farmer in the EC.
b an officer of an aid agency responsible for food supplies to the poorest nations.
Make sure you state the values and attitudes of each person.

Figure 3.15 Energy budgets of average-sized farms in England and Wales

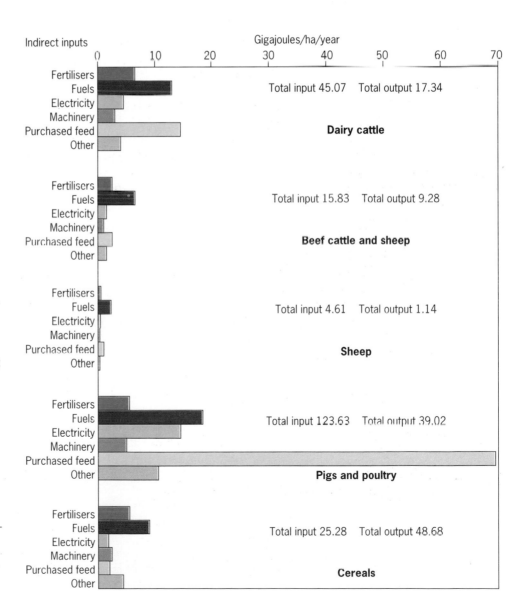

Indirect inputs

Gigajoules/ha/year

Fertilisers
Fuels
Electricity
Machinery
Purchased feed
Other

Total input 45.07   Total output 17.34

Dairy cattle

Fertilisers
Fuels
Electricity
Machinery
Purchased feed
Other

Total input 15.83   Total output 9.28

Beef cattle and sheep

Fertilisers
Fuels
Electricity
Machinery
Purchased feed
Other

Total input 4.61   Total output 1.14

Sheep

Fertilisers
Fuels
Electricity
Machinery
Purchased feed
Other

Total input 123.63   Total output 39.02

Pigs and poultry

Fertilisers
Fuels
Electricity
Machinery
Purchased feed
Other

Total input 25.28   Total output 48.68

Cereals

14a Use the figures in Table 3.6 to draw a similar graph to the one in Figure 3.16.
b Mark on your graph the areas where marginal returns are:
• increasing, • falling, • negative.

15 If you were a farmer aiming to maximise silage production, use Figure 3.16 to decide how much fertiliser per hectare you would apply to your pasture. Give reasons for your decision.

Table 3.6 Potato response to phosphate fertiliser

Phosphate fertiliser (kg per hectare)	0	14	28	56
Potato yield (tonnes per hectare)	13	29	32	32

Limits to inputs: the law of diminishing returns

You might think that the more fertiliser a farmer applies to the crops, the greater the yield. This is true, but only up to a point: the 'law of diminishing returns' tells us that beyond this point the 'value-added' or **marginal return** starts to fall. In other words, each application of fertiliser gives a smaller increase in yield. Eventually, the additional yield is too small to cover the cost of extra fertiliser (Fig. 3.16). This reminds us that there are limits to investment in farming inputs, such as fertiliser and energy.

Figure 3.16 Grass response to nitrogen fertiliser (*Source:* ICI)

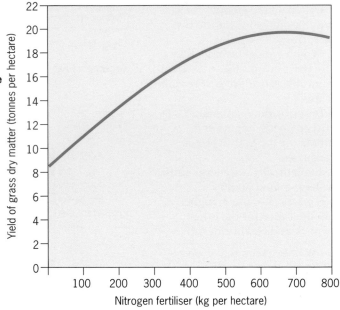

3.9 The system out of balance: chemical pollution

The problem of agricultural pollution, and especially the use of toxic chemicals, has grown rapidly in the second half of the twentieth century. Since 1950, the world use of fertilisers has increased ten times and pesticides by thirty-two times. As yet, we have insufficient knowledge to be certain of the health risks to people. However, the World Health Organisation (WHO) estimates that 200 000 people die each year from pesticide poisoning and that a further three million suffer acute symptoms.

The quality of drinking water contaminated by nitrate fertilisers also gives cause for concern in parts of the UK (Fig. 3.17). In 1990, 1.7 million people in the UK were drinking water containing nitrate levels above the WHO's recommended limit of 50 parts per million (ppm). Excess nitrate in drinking water is linked to blood poisoning in babies, hypertension in children, gastric cancer and birth defects.

Nitrates are highly soluble, so those not taken up by crops percolate slowly into underground aquifers (Figs 3.18 and 3.19). Seepage is often very slow (it can take up to 40 years), so that nitrate contamination is a long-term problem for which there is no simple remedy. Denitrification of water is very expensive. Alternatives include blending contaminated water with other sources of water, setting up protection zones around aquifers where nitrate fertilisers are banned, and changing farming practices.

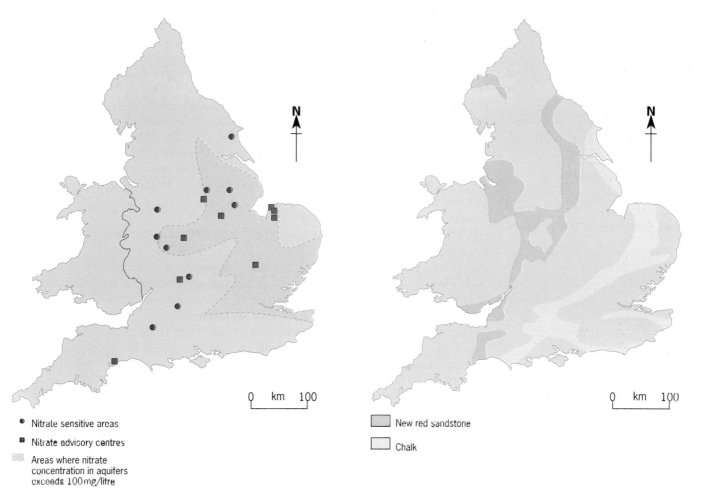

- ● Nitrate sensitive areas
- ■ Nitrate advisory centres
- Areas where nitrate concentration in aquifers exceeds 100 mg/litre

Figure 3.17 Areas of the UK adversely affected by nitrates in groundwater supplies

- New red sandstone
- Chalk

Figure 3.18 Principal aquifers in England and Wales

?

16 Which areas in Figure 3.17 are most likely to suffer health problems from nitrates?

17 Compare the distribution of nitrate contamination in England and Wales (Fig. 3.17) with the maps of principal aquifers (Fig. 3.18) and principal cropping areas (Fig. 4.27). Try to explain the distribution of nitrate-sensitive areas shown in Figure 3.17.

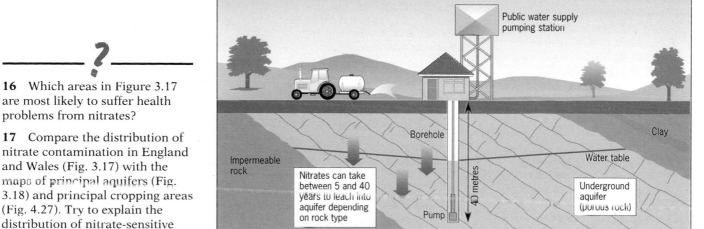

Public water supply pumping station

Borehole

Clay

Impermeable rock

Nitrates can take between 5 and 40 years to leach into aquifer depending on rock type

Water table

40 metres

Underground aquifer (porous rock)

Pump

Figure 3.19 Pollution of groundwater nitrates in south Lincolnshire (*Source: The Daily Telegraph*, 22 Feb. 1988)

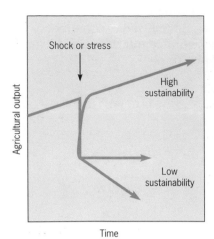

Figure 3.20 The sustainability of farming systems

3.10 Sustainable agriculture

'Sustainable development is a strategy that manages all assets, natural resources, and human resources, as well as financial and physical assets, for increasing *long-term* wealth and well-being. Sustainable development as a goal rejects policies and practices that support *current* living standards by depleting the productive base, including natural resources, and that leave future generations with poorer prospects and greater risks...' (Repetto, 1986).

Rather than exploiting resources so that they are degraded (see Section 6.3), sustainable agriculture seeks long-term development and is willing to sacrifice short-term profit. Figure 3.20 shows that sustainable agriculture is robust. It can absorb environmental shocks such as drought so that production recovers quickly. Agro-ecosystems of low sustainability are, in contrast, fragile, and therefore easily degraded or destroyed.

The concept of sustainable development is a recent one. It was formally introduced by the World Commission on Environment and Development (1987) and was discussed at the UN Conference on Environment and Development at Rio de Janeiro in 1992 (Table 3.7).

Alternative farming: sustainable agriculture in economically developed countries

We saw in Sections 3.2–3.4 that modern commercial agriculture is intensive and relies on large inputs of energy. **Alternative farming** is more **extensive** and more environmentally friendly. It takes three forms: diversified farming; low input-output farming; and organic farming.

Diversified farming is based on a variety of crops (including timber) and/or livestock, and includes crop rotations. Low input-output farming uses reduced amounts of energy and fertiliser, leading to less intensive farming methods and lower production. The result is less damaging to the environment. Organic farming removes agrochemicals and inorganic fertilisers from farm inputs, and maintains soil fertility by using organic waste.

Table 3.7 The Rio conference: main decisions taken

- To adopt the Rio Declaration on Environment and Development with its 27 principles on the duties and rights of individual countries and the establishment of a global consensus on sustainable development.
- To produce a plan of action for the twenty-first century (Agenda 21) dealing with the conservation and management of resources for development.
- To set up an international Framework Convention on Climatic Change.
- To set up a Framework Convention on the Conservation of Biological Diversity, signed by 154 countries.
- To make a statement of principles for the management, conservation and sustainable development of all types of forest resources.
- To start negotiations for an international convention to combat desertification.

Table 3.8 The techniques of organic farming (*Source*: Marland, 1989)

1 On-farm waste recycling of manure and crop residues. This is seen to be more in harmony with natural ecosystems than the use of chemical fertilisers and the burning of straw.

2 Non-manufactured, naturally occurring mineral fertilisers (e.g. limestone) are allowed, but only where nutrients are released through weathering or the activity of soil organisms.

3 Weeds are controlled by:
   - mulching (when a layer of organic material, e.g. straw, is applied to the soil surface to reduce evaporation and stop weed growth),
   - crop rotation,
   - cultivation rather than spraying,
   - timed planting to allow the weed seeds to germinate first and emerge before the crop.

4 Pests and diseases are prevented by:
   - using less nitrogen thus allowing plants to become sturdier and less succulent to pests,
   - rotating crops to reduce the chance of a pest surviving from one year to the next,
   - encouraging predators e.g. spiders, birds,
   - using natural biocides like pyrethrum and derris,
   - using resistant crop varieties.

5 Animal welfare is given a high priority.

6 The use of green manuring and under-sown legumes which fix nitrogen in the soil. Some farmers aim for green cover all year round.

7 Crop rotations, such as:
   - a four year grass/clover ley, then barley, oats, beans, oats,
   - a four year ley, then wheat, barley and undersown clover, ley, wheat, oats,
   - a two year ley, then wheat, barley, potatoes, wheat, oats,
   - beans/peas, then wheat, oats, beans, wheat, rye.

8 Inter-cropping, or the mixing of rows of different crops in the same field. This reduces the risks of pests spreading from one plant to another in the same crop, and makes better use of sunlight.

Organic farming

For our purposes, we can define organic farming as 'farming without factory-made chemicals, either as fertilisers, pesticides, herbicides or yield enhancing drugs for animals'. There are about 10 000 organic farmers in Europe, including 1000 in the UK. In some respects organic farming is a return to traditional methods which were far less intensive than modern agriculture (Table 3.8). It is kinder to the environment because fewer toxins (from chemical fertilisers, pesticides etc.) are released into the soil or leached into streams and rivers. It also uses less energy and is therefore more likely to conserve the earth's fossil fuel resources.

In the UK, organic farmers are usually owner-occupiers of smaller than average holdings, and only about one quarter of these have always been farmers. They tend to be concentrated in the south and south-west of England and in west Wales. In 1987, it was estimated that around 55 000 hectares were cultivated organically, which represents only 0.3 per cent of all UK farmland.

Organic farming tends to be practised by people who have a deep commitment to the environment. Yet the growth of organic farming is also consumer-led, because many people are worried about the chemical residues in the food they buy. People are therefore willing to switch to food produced in what they regard as a healthier system. Thus a market for organic food has developed, even though prices are higher than food produced by conventional farming methods. None the less, in 1990 sales of organic food only accounted for 0.25 per cent of the UK's grocery bill.

Apart from the extra cost of organically grown food, organic farming has a further drawback: its yields are 10–20 per cent lower than conventional farming. This is due to greater losses to pests and diseases, to lower soil fertility and less capital-intensive inputs. Reduction in output is, however, compensated by higher prices (Table 3.9).

In spite of the favourable figures shown in Table 3.9, the profitability of organic farming is as yet unproven, especially for certain products. Sheep farming is particularly problematic, because sheep easily pick up parasites and it is difficult to control these without chemicals. Understandably, many farmers are cautious about converting to organic systems (Fig. 3.21).

?

18 Study the figures in Table 3.9.
a By what percentage do organic yields and market price for winter wheat differ from conventional yields and market price?
b From the evidence of Table 3.9, explain why winter wheat is more profitable using organic rather than conventional farming methods.

Table 3.9 Comparison of profitability of organic and conventional wheat growing (*Source*: Organic Farmers and Growers Ltd, 1992)

Organic system		Conventional system	
	Winter wheat £ per hectare		Winter wheat £ per hectare
Output: Yield 2 tonnes at £518.90	1037.80	Output: Yield 3 tonnes at £313.80	941.50
Inputs:		Inputs:	
Seed	69.20	Seed	59.30
Minerals	44.50	Compound	59.00
Seaweed, sulphur	24.70	Nitrogen	40.80
Manures	37.10	Herbicide	34.60
Organic N fertiliser	74.10	Slug pellets, aphicides	24.70
Extras	24.70	Fungicides	24.70
Total variable costs	274.30	Total variable costs	258.80
Net profit per hectare	763.50	Net profit per hectare	682.70

?

19 Read the article in Figure 3.21.

a Describe what converting to organic farming would involve for the Goodmans.

b Draw an annotated diagram to show how farm inputs would change if the Goodmans switched to organic farming.

c Although outputs fall under organic farming, profits should rise. Explain this paradox.

d Describe the major obstacles which the Goodmans would face if they switched to organic farming.

20 Look ahead to Chapter 12.

a Consider the views presented in Table 12.1.

b Write a dialogue between an environmentalist and a technologist discussing organic farming. Express the attitudes of both values positions.

c What is your own attitude to organic farming? Does it have a successful future? Express your opinion, giving reasons.

21 Essay: Compare organic farming with **agribusiness** (see Section 8.9). Use systems diagrams to support your answer.

Farmer counts the cost of going organic

Michael Hornsby

Geoff Goodman, who owns and rents 881 hectares of pasture and arable land, runs the farm with his wife Judy, son Andrew and one full-time employee.

Earlier this year, the family called in a consultant to advise on the feasibility of converting to organic methods. 'The notion that you can reduce output, so putting less pressure on land and livestock, and make a better profit than before, is seductive and we had been thinking about the organic option for some time,' Mr Goodman said.

At present the Goodmans have a dairy herd of 145 cows, 63 beef cattle, 96 hectares of permanent and temporary pasture and 49 hectares of wheat, barley and fodder crops. Mrs Goodman also runs a free-range flock of 1 800 geese.

Under the consultant's blueprint, the Goodmans would have concentrated on producing milk and cereals. The beef cattle would have gone, leaving more pasture for the dairy herd. The land now under barley and fodder beet would have been turned over to wheat and oats. Only farmyard manure would have been permitted as fertiliser. To maintain fertility, the Goodmans would have to rely on crop rotation, which in turn would have entailed re-seeding up to 81 hectares with clover rich grass to 'fix' nitrogen in the soil. No chemical pesticides would have been allowed.

At least 80 per cent of the cows' feed would have had to be organically grown. Veterinary drugs would have been allowed if animals fell ill, but there would have been a ban on routine use of antibiotics to prevent infections.

Cereal yields, it was calculated, would have fallen from 3 to 1.8 tonnes and milk yields from 6 000 to 55 300 litres per cow. This fall in output however would have been more than offset by the assumed higher price of organic produce — 22.2p a litre for milk (compared with 18.3p for non-organic) and £220 a tonne for cereals (as against £112).

Money would have been saved on fertilisers, chemicals and drugs, but feed-stuffs, fuel and machinery maintenance would have cost more. A second full-time farmhand might have been needed. Still, the bottom line looked quite good: once the farm was running as an organic unit, the family could expect a net income of £29 700 a year, £10 000 more than now. The snag was how to get there.

The Goodmans were told they would need to spend £40 000 on re-seeding pasture, on extra storage for grain and manure and on new machinery for spreading muck and weeding un-sprayed fields. More worrying was the estimate that full organic status might take five years to achieve. During that time, farm output would be falling without the compensation of higher prices.

'We simply did not see how we could contemplate these risks and costs, particularly when we are already paying £1 7850 a year in rent and interest on bank loans. I was also not persuaded that the market premium for organic produce would always be there. Frankly, milk is about as natural a commodity as you can get, however you produce it.'

Figure 3.21 One farmer tries the sustainable route (*Source: The Times*, 21 Oct. 1991)
© Times Newspapers Ltd 1991

Summary

- Agriculture is a biological system (agro-ecosystem), comprising crops, livestock, soil, water, gases and flows of energy and nutrients.

- Agro-ecosystems are manipulated by farmers to maximise useful (i.e. food) production.

- Like all ecosystems, agro-ecosystems are inefficient in their conversion of energy. Crops capture only a tiny fraction of sunlight which they convert into chemical energy through photosynthesis. Even less efficient is the conversion of plant energy to animal tissue.

- Commercial agriculture is highly productive, though its high yields are only sustained by importing energy (such as fuel for machinery) and nutrients (in the form of chemical fertiliser) from outside the farm system.

- Agricultural output does not rise proportionately with inputs: eventually the law of diminishing returns operates, making further inputs unprofitable.

- The use of toxic chemicals in agro-ecosystems can be damaging to the environment. Nitrates have caused particular concern in the UK.

- In the economically developed world, concern about the harmful effects of commercial agriculture has led to an interest in more traditional farming methods and the idea of sustainability.

4 Climatic limits to crop growth

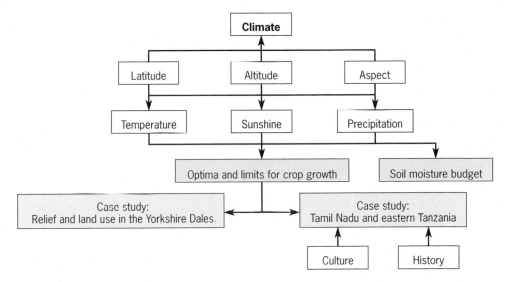

```
                          ┌──────────┐
                          │ Climate  │
                          └──────────┘
        ┌────────────┬──────────┴──────┬─────────────┐
   ┌─────────┐  ┌──────────┐      ┌──────────┐
   │ Latitude│  │ Altitude │      │  Aspect  │
   └─────────┘  └──────────┘      └──────────┘
        │            │                 │
   ┌───────────┐ ┌──────────┐    ┌──────────────┐
   │Temperature│ │ Sunshine │    │ Precipitation│
   └───────────┘ └──────────┘    └──────────────┘
              │        │                │
      ┌───────────────────────────┐  ┌──────────────────┐
      │ Optima and limits for     │  │ Soil moisture    │
      │ crop growth               │  │ budget           │
      └───────────────────────────┘  └──────────────────┘
   ┌─────────────────────────┐   ┌─────────────────────────┐
   │ Case study:             │◄─►│ Case study:             │
   │ Relief and land use in  │   │ Tamil Nadu and eastern  │
   │ the Yorkshire Dales     │   │ Tanzania                │
   └─────────────────────────┘   └─────────────────────────┘
                                    ┌─────────┐  ┌─────────┐
                                    │ Culture │  │ History │
                                    └─────────┘  └─────────┘
```

1 Study the landscape in Figure 4.1

a List the physical factors that are likely to have influenced farming.

b For each factor, suggest how, and to what extent, farmers can modify it to grow crops.

4.1 Introduction

Despite impressive technological advances, agriculture relies heavily on the physical environment. Crops require specific temperature, moisture, soil and drainage conditions which farmers can modify, but only to a limited degree. The limits are set by money and available technology, but especially by the physical system itself. As a result, the distribution of agricultural activity is largely controlled by the physical resources on which it depends.

Warmth, light and **precipitation** are physical inputs that set limits on the crops which farmers can grow. Coffee (Fig. 4.2), for example, is frost-sensitive and will not grow where mean monthly temperatures fall below 11°C, and where mean annual precipitation is outside the range of 200–900 millimetres. On the other hand, rice (front cover), which comprises many different varieties, is adapted to a wider range of climatic conditions. Although most rice production is found within the tropics, cultivation extends over a vast area, from latitude 45°N to 40°S.

Figure 4.1 Littondale, Yorkshire Dales, UK

Figure 4.2 Harvesting coffee, Cameroon

4.2 The influence of latitude and temperature

At the global scale, solar radiation is the major influence on temperatures (Fig. 4.3).

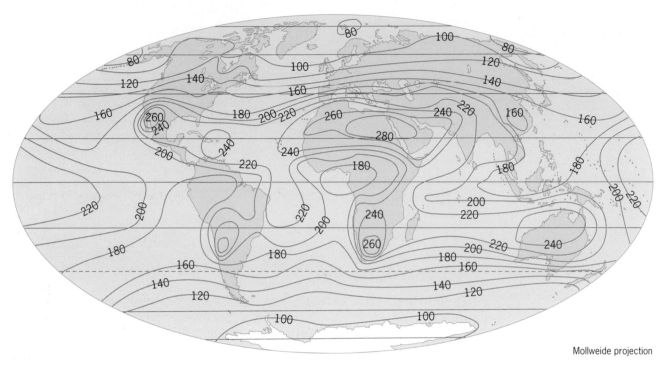

Mollweide projection

Figure 4.3 Average annual solar radiation (watts per m²) on a horizontal surface at ground level (*Source:* Barry and Chorley, 1968)

Temperature

When solar radiation strikes the land surface it warms the ground and this warmth is then transferred to the overlying air. All crops require minimum amounts of heat to complete their growth cycles from germination to fruiting. However, there is no precise temperature at which crops start to grow, though, as a general rule, we often assume that growth begins when the mean daily air (or soil) temperature reaches 6°C. We refer to the number of days each year when temperatures exceed this limit as the **thermal growing season**. At sea level in Europe, this varies from 365 days in southern Spain, to 150 days in central Norway.

An alternative measure, which gives some idea of the warmth of the growing season, is **accumulated temperatures**. These are calculated by adding together the mean daily temperatures above a chosen limit (usually 6°C for middle to high latitudes), and expressing them as the number of **degree days**.

Outside the tropics, the thermal growing season is the main physical limit on agriculture. But for frost-sensitive crops like cotton or coffee, the number of days when temperatures are above freezing is more important.

Latitude

Latitude describes the position of a place north or south of the Equator. It affects the growing season, and this is seen most at a global scale. Yet, even within a small area such as England, latitude influences the length of the growing season (Figs 4.5–4.6).

2 Study Figure 4.3.
a Comment on the distribution of radiation intensity with regard to the equator, the tropics and the poles.
b Give reasons for this global pattern.

3 Using Figure 4.4, predict the average length of the growing season in England at latitudes 51°N, 52°N, 53°N.

4 What factors other than latitude might influence the growing season in Figure 4.4?

5 Look at Figure 4.5 and find the length of the growing season at the East Durham farm described in Section 2.4.

6 Study carefully the growing season in England and Wales as measured by:

a the number of days with temperatures 6°C and above (Fig 4.5).

b the number of degree days above 10°C (Fig. 4.6). Comment on the validity of the following null hypothesis:

'Farmers in South-East England have no climatic advantage over farmers in South-West England'.

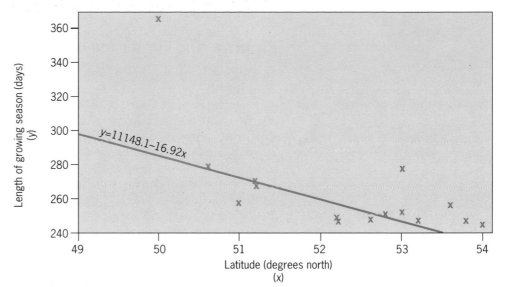

$y = 11148.1 - 16.92x$

Figure 4.4 Latitude and length of growing season in England for places below 65m above sea-level

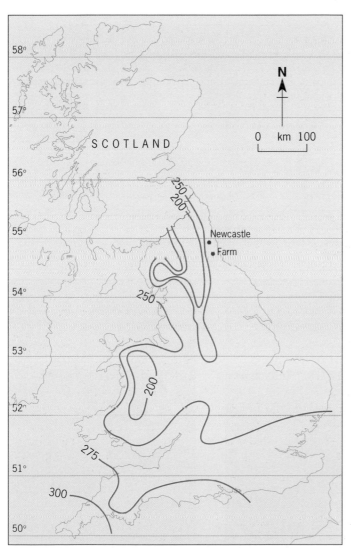

Figure 4.5 Length of growing season in England and Wales: number of days when soil temperatures (at 30cm depth) exceed 6 °C

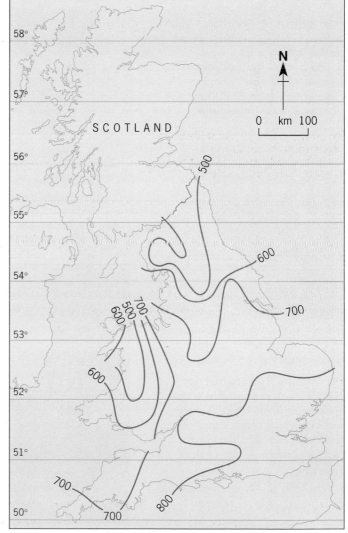

Figure 4.6 Length of growing season in England and Wales: number of degree days above 10 °C, May–October

Table 4.1 Critical temperature ranges for the growth of cereal crops

	Temperate cereals (e.g. wheat, barley, oats) °C	Tropical/sub-tropical cereals (e.g. rice, maize, millet) °C
Minimum	0–5	15–18
Optimum	25–31	31–37
Maximum	31–37	44–50

7 Study Figure 4.7.
a What climatic factors determine: • the western, • the northern, • the southern limits of cotton cultivation?
b Within the cultivation area defined by climate, where is cultivation optimal? Define the climatic limits of this area.

8 Figure 2.11 shows vast areas where climatic conditions are too severe for agriculture. Using evidence from Figures 4.8 and 4.9 (which show extreme physical environments) suggest reasons why agriculture is absent from these places.

9 Study Figures 4.10–4.12.
a Using Figure 4.10, describe the distribution of wheat and rice.
b With reference to Figures 4.10–4.12, compare the actual temperatures under which cereal crops grow with the optimal temperatures given in Table 4.1.

10 Neither the minimal nor optimal temperatures for crops are permanently fixed. Suggest ways in which science and technology could alter the temperature limits for crop growth.

Optima-and-limits for crop growth

In Table 4.1, the minimum temperatures indicate that there is insufficient heat for biological activity. For example rice, maize and millet need temperatures of at least 15–18°C throughout the growing season. In northern Europe, low temperatures do not allow for cultivation of these cereal crops.
Optimal temperatures (Table 4.1) give the best thermal conditions for crop growth. However, if temperatures are too high, then growth slows down. Eventually, if temperatures rise above the maximum, growth stops altogether.

We can explain the distribution of crop production in terms of simple optima-and-limits models (Fig. 4.7). We begin by stating that crops can only be grown successfully in areas which meet the crops' minimal requirements for heat, moisture and soils. Because many areas do not provide these requirements, physical factors set spatial limits to crop production. However, within the area that does meet the minimal requirements, there are smaller areas where combinations of physical factors may give optimal conditions for crop growth.

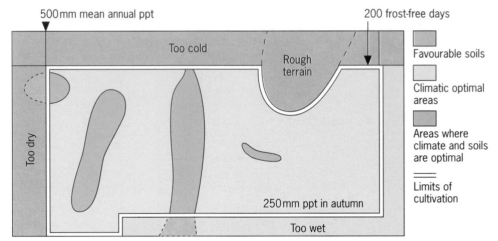

Figure 4.7 Optima-and-limits model for cotton-growing in the southern USA (*Source:* MacCarty and Lindberg)

Figure 4.8 Victoria Island, Canada

Figure 4.9 Sahara Desert, northern Africa

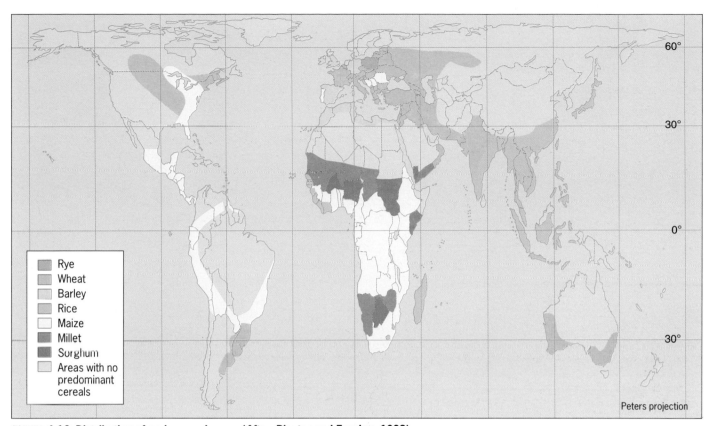

Figure 4.10 Distribution of main cereal crops (*After:* Blaxter and Fowdon, 1982)

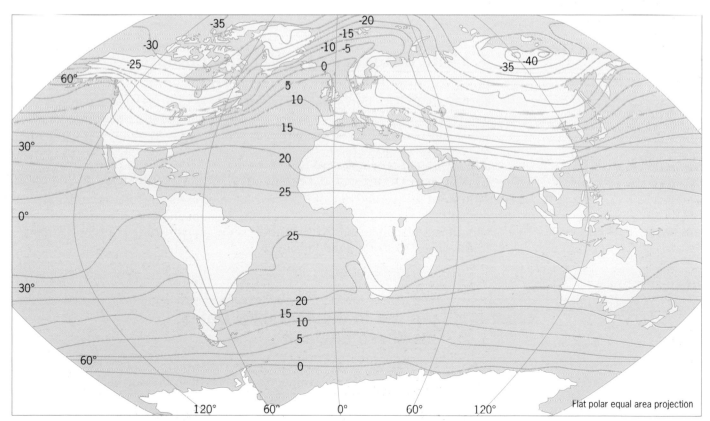

Figure 4.11 Mean sea-level temperatures (°C), January

Optima-and-limits for crop growth

Figure 4.12 Mean sea-level temperatures (°C), July

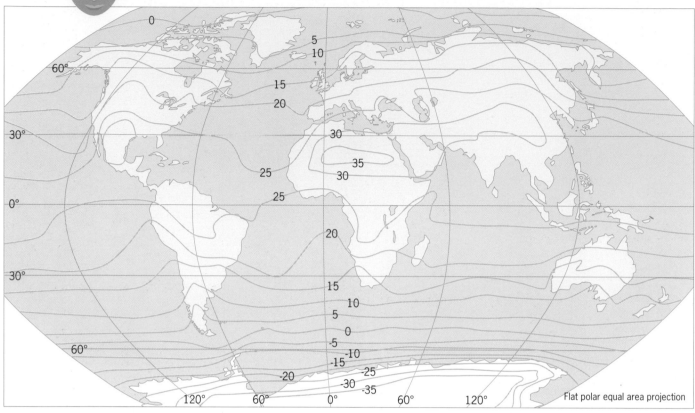

Flat polar equal area projection

4.3 The influence of relief and climate

Relief describes the altitude (or height) and the slope (or gradient) of the land surface. Both have important influences on agriculture, especially at regional and local scales. Altitude is a major control on temperature and precipitation, while slopes also affect temperature, as well as the workability of farmland.

Altitude and temperature
In the lowest 15 kilometres of the atmosphere, temperatures decrease with altitude at an average rate of 6.5°C per kilometre. As a result, upland areas are both cooler and have shorter growing seasons than the lowlands. Cloudiness also increases with altitude, reducing sunshine levels and adding to the difficulties of upland farming.

Relief and land use in the Yorkshire Dales, northern England

If we look at a world map of agriculture (Fig. 2.11), it is clear that mountainous areas such as the Himalayas, Andes and Alps support little agricultural activity. Even in the British Isles, the fall in temperature with altitude means that uplands above 350 metres are too cold and too cloudy for cultivation. At this height in the northern Pennines and the Lake District, the growing season lasts on average for just 200 days a year.

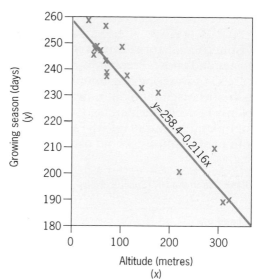

Figure 4.13 Altitude and the length of the growing season in northern England

$y = 258.4 - 0.2116x$

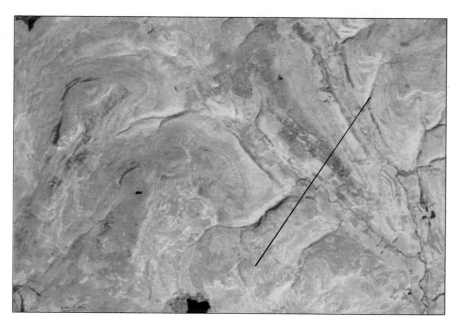

Figure 4.14 Landsat image of Littondale area, Yorkshire Dales
Orange = open grassland; olive green = upland moors; brown = conifer woodland; blue = limestone outcrops; black = water

11 Using the best-fit (regression) line in Figure 4.13, calculate the average decrease in the length of the growing season in northern England for every 100 metres increase in altitude.

12 Study Figures 4.14, 4.15 and 4.16.
a Draw a relief section from Figure 4.15 between Parson's Pulpit (919688) and Starbotton (954738).
b The line of the section you have drawn is shown on the satellite image (Fig. 4.14). Using the satellite image and the colour key, mark on to your section the main types of land use.
c Describe and explain how land use is influenced by altitude in the Yorkshire Dales.
d What other factors might affect land use in this region (Fig. 4.16)?

Figure 4.16 Sheep farming at Malham Cove, Yorkshire Dales, UK

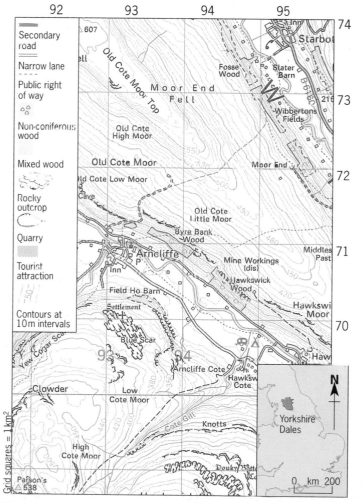

Figure 4.15 The Yorkshire Dales: (© Crown Copyright, 1989)

Altitude and precipitation

Precipitation describes all types of atmospheric moisture. Thus it includes not only rain but also drizzle, fog, hail, sleet and snow. Precipitation increases with altitude and is often the cause of waterlogging in upland soils. Heavy precipitation may also **leach** (wash out) essential plant nutrients from upland soils, adding further to the problems faced by farmers. Not surprisingly, the agricultural potential of upland areas outside the tropics is strictly limited. The effects of altitude and precipitation on land use in the UK are summarised in Table 4.2.

Table 4.2 Altitude, precipitation and agricultural land use in the UK (*Source*: Bibby and Mackney, 1969)

Altitude and mean annual precipitation	Land use
Land over 600 m.	Above the tree-line; poor, rough grazing only.
Land 300–600 m with over 1500 mm precipitation.	Rough grazing; pasture improvement is usually not possible.
Land 200–300 m with under 12 500 mm precipitation.	Improved pasture but also suitable for crops.
Land 100–200 m with over 1000 mm precipitation.	Mainly suitable for improved grass and limited arable crops.

Slopes and temperature

While latitude controls how much solar radiation a place receives at a global scale, at a local scale **aspect** and slope are important influences. Aspect refers to the direction in which a slope faces, so that in middle to high latitudes, the sun-facing or **adret** slopes have warmer micro-climates than shadowed or **ubac** slopes. This is most evident in deep valleys incised into upland areas. For example, while there are many vineyards on the south-facing adret slopes of the Rhine and Mosel valleys in central Germany, the colder ubac slopes are often uncultivated and forest-covered (Fig. 4.17).

The intensity of solar radiation is greatest where the sun's rays strike the ground at right angles. Outside the tropics this only happens on slopes which face the sun. Consequently, these adret slopes offer a significant advantage to agriculture: a south-facing slope of 20 degrees in Europe receives an intensity of solar radiation equal to a southerly shift of 8 or 9 degrees of latitude. But a similar north-facing slope has a huge disadvantage: in this case it is equivalent to a northerly shift in latitude of 12 to 15 degrees! (Table 4.3)

Cold-air drainage also affects agriculture in deep valleys (Fig. 4.18). At night, air over slopes cools more quickly than air at the same height above a valley. This allows cold dense air to creep downslope and accumulate on the valley floor (the **frost hollow**). As the air stagnates here, temperatures often fall well below freezing.

Figure 4.17 Adret and ubac slopes with viticulture in the Mosel Valley, Germany

Table 4.3 Potential radiation income at Kinlochleven, Scotland, 57°N

	Noon Mid-summer (watts per m²)	Mid-winter (watts per m²)
25° adret slope	290	3.33
>30° ubac slope	103	Shadowed
Valley floor	235	Shadowed

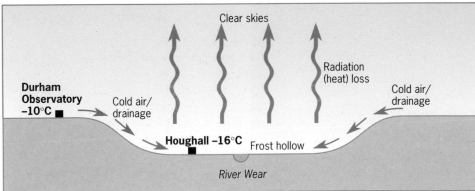

Figure 4.18 Cold air drainage and temperature inversion at night, Wear Valley, Co. Durham (February, 1941)

13 Fruit trees are sensitive to frost. A friend wishes to plant an apple orchard in Houghall (Fig. 4.18). Advise your friend what to do, and give reasons for your advice.

Slopes and workability

Steep slopes create problems for farm machinery. For instance, two-wheel drive tractors can only operate on slopes of up to 15 degrees, and four-wheel drive tractors to maximum slope angles of 25 degrees. Above 30 degrees, farmers cannot use vehicular machines at all. A further problem on steep slopes is the risk of soil erosion caused by runoff. For these reasons farmers do not normally grow crops on slopes of more than 10 degrees.

4.4 The influence of precipitation

In the tropics, the thermal growing season lasts all year round. Thus soil moisture, rather than temperature, sets the limit on crop and livestock production in these areas. However, the **hydrological growing season** (the period when there is sufficient moisture for crop growth) depends more on **precipitation effectiveness** than on how much precipitation falls. Precipitation effectiveness is the amount of precipitation available to plants after you deduct losses due to evaporation and transpiration. The two major factors which determine precipitation effectiveness are: a) precipitation characteristics, and b) **evaporation** and **transpiration** (Fig. 4.19).

Figure 4.19 Factors influencing the effectiveness of precipitation for crop growth

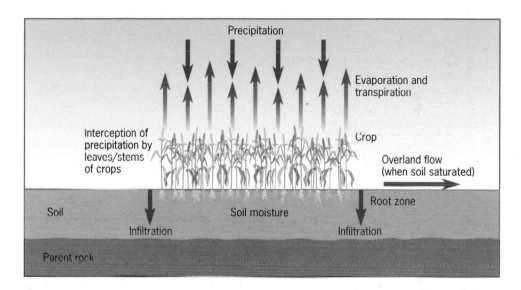

Table 4.4 Precipitation and evaporation at Shrewsbury and Buxton

	Shrewsbury (altitude 100 m)	Buxton (altitude 305 m)
Mean annual precipitation (mm)	575	1200
Mean annual evaporation (mm)	475	375

14a Using the information in Table 4.4, calculate the effective precipitation at Shrewsbury and Buxton.
b The amount of precipitation at Buxton is approximately double that at Shrewsbury. How many times greater is the effective precipitation at Buxton? Try to explain this difference, using Figure 4.19.
c Draw an annotated diagram to show how the difference in effective precipitation is likely to affect soils and their value to agriculture.

Precipitation characteristics

In many parts of the world farmers are more interested in the **seasonal distribution** of precipitation rather than in the amount of precipitation (see the case study of Tamil Nadu and eastern Tanzania at the end of this chapter). For example, in the Mediterranean most precipitation falls during the cooler part of the year (Fig. 4.20). Losses by evaporation are therefore reduced and so a higher amount of the total precipitation is available as water for crops. In higher latitudes, such as the Canadian prairies, a summer maximum of precipitation must coincide with the growing season to produce a successful wheat harvest (Fig. 4.21).

?

15 Look at Figures 4.20–4.21.
a Compare the two climates and describe the differences in precipitation amounts and times.
b Give reasons for these.

16a Compare Figures 4.22 and 4.23 Comment on the hypothesis that, as mean annual precipitation increases, precipitation variability increases in Africa.
b Explain why dryland farming is absent at locations A and B in Figure 4.22.

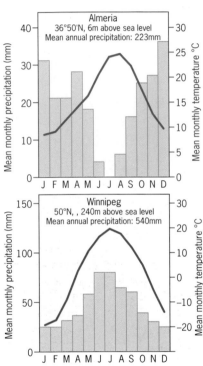

Figures 4.20–4.21 Climate in Almeria, Spain (*above*) and Winnipeg, Canada (*below*)

Precipitation intensity also affects cultivation. Thus, where precipitation takes place during high-intensity storms which cause rapid runoff, little water is available to crops. In contrast, low-intensity precipitation spread over several hours will allow water to infiltrate the soil and reach the root zone of crops (Fig. 4.19).

Finally, agriculture is affected by the year-to-year changes in precipitation (Figs 4.22 and 4.23). Typically, in arid and semi-arid regions, mean annual precipitation may vary by as much as 50 per cent. Because of this uncertainty, it is not possible for a permanent dryland farmer to survive. Thus, in Africa and the Middle East, the margin of dryland cultivation matches the 250 millimetre isohyet *and* an average annual variability of less than 40 per cent.

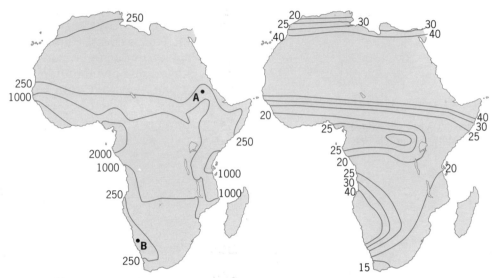

Figure 4.22 Africa: mean annual precipitation (mm)

Figure 4.23 Africa: variability of precipitation (per cent)

Evaporation

We have seen that precipitation effectiveness depends largely on evaporation. Moisture losses through evaporation are greater in warm than in cold climates, and are higher in summer than in winter. This relationship between precipitation and evaporation has a strong influence on moisture levels in the soil. We describe this relationship over a year by using the idea of a **soil moisture budget**.

Soil moisture budgets

The soil moisture budget (Figs 4.24 and 4.25) is very important for farmers. A long dry period, or moisture deficit, in summer will reduce crop growth and give poorer yields. Grass, for example, stops growing when the moisture deficit in the root zone is greater than 50 millimetres. The date on which the soil again contains maximum moisture, or returns to **field capacity** is also of great importance to farmers. After this date, the soil becomes saturated and is difficult to work with machinery without causing permanent damage. It is partly for this reason that cultivation throughout northern Europe comes to a virtual standstill in late autumn.

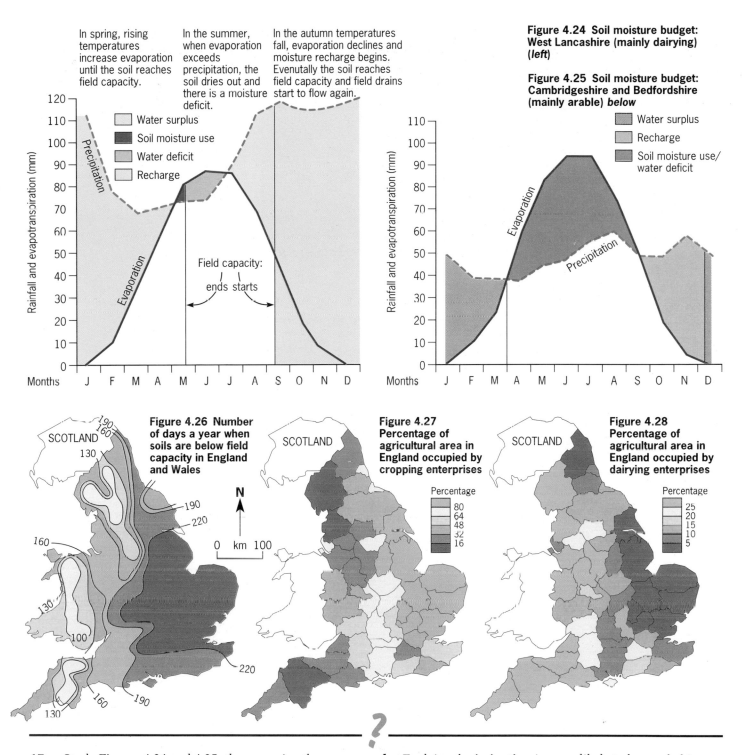

In spring, rising temperatures increase evaporation until the soil reaches field capacity.

In the summer, when evaporation exceeds precipitation, the soil dries out and there is a moisture deficit.

In the autumn temperatures fall, evaporation declines and moisture recharge begins. Evenutally the soil reaches field capacity and field drains start to flow again.

Figure 4.24 Soil moisture budget: West Lancashire (mainly dairying) (left)

Figure 4.25 Soil moisture budget: Cambridgeshire and Bedfordshire (mainly arable) below

Figure 4.24 legend:
- Water surplus
- Soil moisture use
- Water deficit
- Recharge

Figure 4.25 legend:
- Water surplus
- Recharge
- Soil moisture use/ water deficit

Figure 4.26 Number of days a year when soils are below field capacity in England and Wales

Figure 4.27 Percentage of agricultural area in England occupied by cropping enterprises

Percentage
80
64
48
32
16

Figure 4.28 Percentage of agricultural area in England occupied by dairying enterprises

Percentage
25
20
15
10
5

17a Study Figures 4.24 and 4.25, then examine the following statements and say whether they are true or false:

- Precipitation is higher in West Lancashire.
- Evaporation is the same in West Lancashire and Cambridgeshire and Bedfordshire.
- West Lancashire has a longer period of moisture deficit.
- The soil is at field capacity for longer in Cambridgeshire and Bedfordshire.

b Explain why **irrigation** is more likely to be needed in Cambridgeshire and Bedfordshire, and why the land is workable for a shorter period in West Lancashire.

18 Explain how the pattern of soil moisture deficit in England and Wales (Fig. 4.26) might influence a farmer's choice of cropping or dairy enterprises (Figs 4.27 and 4.28). Choose two contrasting locations to illustrate your answer.

Physical factors and farming in Tamil Nadu and eastern Tanzania

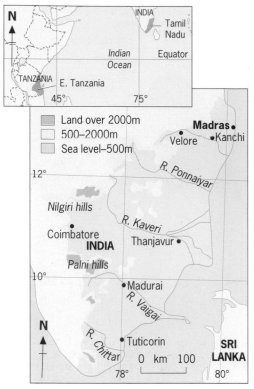

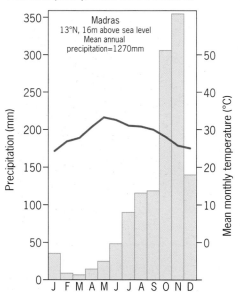

Figure 4.29–4.30 Tamil Nadu and eastern Tanzania (*inset*); Tamil Nadu: features

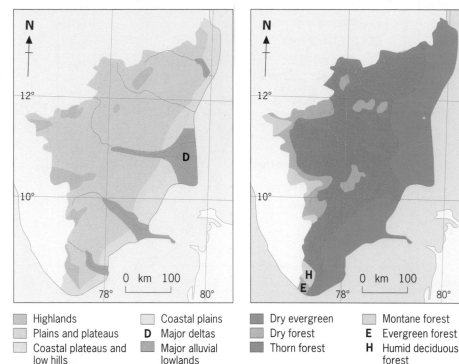

Figure 4.31 Tamil Nadu: relief
(*Source* of maps and data on pp 50–1: Morgan, 1988)

Highlands | Coastal plains
Plains and plateaus | **D** Major deltas
Coastal plateaus and low hills | Major alluvial lowlands

Figure 4.32 Tamil Nadu: vegetation

Dry evergreen | Montane forest
Dry forest | **E** Evergreen forest
Thorn forest | **H** Humid deciduous forest

Figure 4.33 Tamil Nadu: climate

Madras
13°N, 16m above sea level
Mean annual
precipitation=1270mm

Figure 4.34 Paddy fields in Tamil Nadu, irrigated from the reservoir at right of the photograph

Physical factors

In Tamil Nadu in southern India, and in eastern Tanzania in East Africa (Figs 4.29, 4.30, 4.35), physical factors have an important influence on agriculture.

In terms of physical geography both regions are similar, and physical resources (notably relief, soils and climate) determine the scope for crops and farming types.

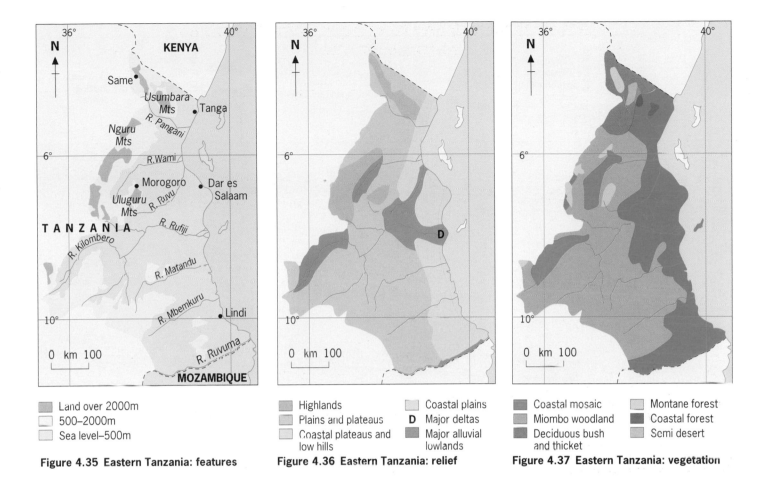

Land over 2000m
500–2000m
Sea level–500m

Figure 4.35 Eastern Tanzania: features

Highlands
Plains and plateaus
Coastal plateaus and low hills
Coastal plains
D Major deltas
Major alluvial lowlands

Figure 4.36 Eastern Tanzania: relief

Coastal mosaic
Miombo woodland
Deciduous bush and thicket
Montane forest
Coastal forest
Semi desert

Figure 4.37 Eastern Tanzania: vegetation

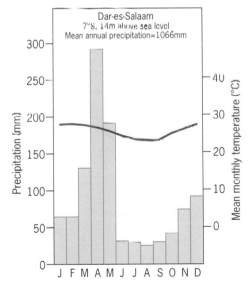

Figure 4.38 Eastern Tanzania: climate

Figure 4.39 Farming land near Malawi, Eastern Tanzania

Relief

From Figures 4.31 and 4.36 you will see that both Tamil Nadu and eastern Tanzania have extensive plains above the coastal regions. These are based on crystalline rocks, and **inselbergs**, or isolated hills, rise steeply above the plains. Larger hill ranges then rise to 2000 metres in altitude. At lower levels in both regions, there are shallow, seasonally wet valleys with better drained **interfluves** between them.

Tamil Nadu and eastern Tanzania

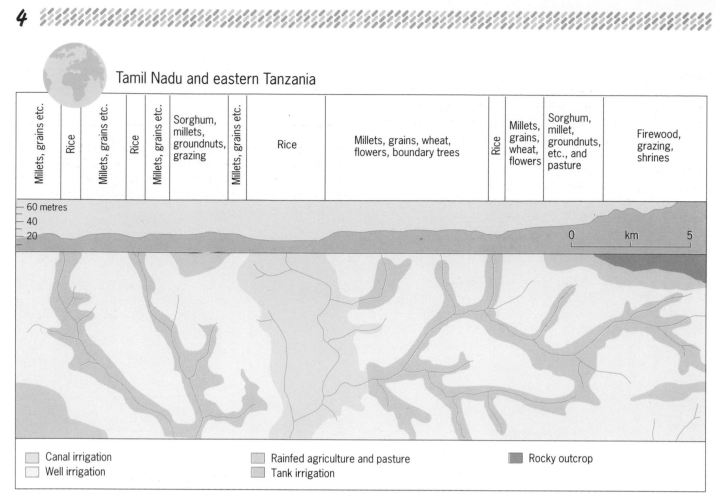

Millets, grains etc.	Rice	Millets, grains etc.	Rice	Millets, grains etc.	Sorghum, millets, groundnuts, grazing	Millets, grains etc.	Rice	Millets, grains, wheat, flowers, boundary trees	Rice	Millets, grains, wheat, flowers	Sorghum, millet, groundnuts, etc., and pasture	Firewood, grazing, shrines

— 60 metres
— 40
— 20

0   km   5

☐ Canal irrigation   ☐ Rainfed agriculture and pasture   ☐ Rocky outcrop
☐ Well irrigation   ☐ Tank irrigation

Figure 4.40 Tamil Nadu: idealised land use

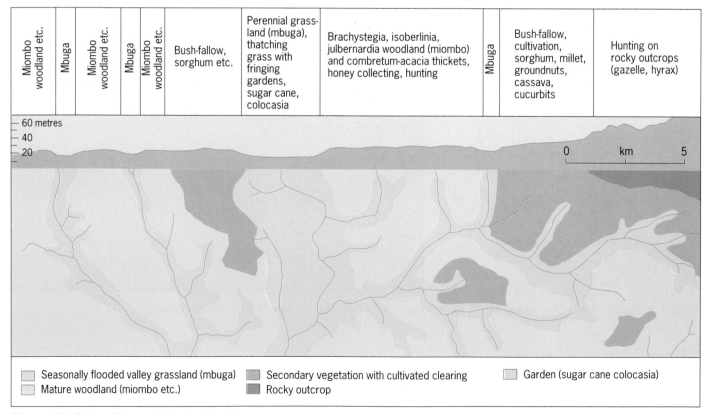

Miombo woodland etc.	Mbuga	Miombo woodland etc.	Mbuga	Miombo woodland etc.	Bush-fallow, sorghum etc.	Perennial grass-land (mbuga), thatching grass with fringing gardens, sugar cane, colocasia	Brachystegia, isoberlinia, julbernardia woodland (miombo) and combretum-acacia thickets, honey collecting, hunting	Mbuga	Bush-fallow, cultivation, sorghum, millet, groundnuts, cassava, cucurbits	Hunting on rocky outcrops (gazelle, hyrax)

— 60 metres
— 40
— 20

0   km   5

☐ Seasonally flooded valley grassland (mbuga)   ☐ Secondary vegetation with cultivated clearing   ☐ Garden (sugar cane colocasia)
☐ Mature woodland (miombo etc.)   ☐ Rocky outcrop

Figure 4.41 Eastern Tanzania: idealised land use

Climate

Tamil Nadu has a higher mean annual temperature than eastern Tanzania but only by 2°C. Mean annual precipitation is also at similar levels and in both cases this decreases inland (Figs 4.33 and 4.38).

Vegetation

Natural vegetation varies from dense forest to woodland with variable tree cover (Figs 4.32 and 4.37). However, the human impact on the landscape of Tamil Nadu has been considerable and little natural vegetation cover survives today.

People adapting to the physical environment

As we can see from Figures 4.40 and 4.41, the human response to the physical resource base has been very different in these two regions. This reflects the influence of human factors, especially the contrasts in history and culture between southern India and East Africa.

19a Use all the information in Figures 4.30–4.39 to compare the physical environment of Tamil Nadu and eastern Tanzania. Present your comparison as a table under the following headings: latitude and area; climate; relief; vegetation.
b Describe how physical factors have influenced land use in Tamil Nadu and eastern Tanzania (Figs 4.40 and 4.41).
c Say whether you would have expected these patterns from the factors shown in Figure 4.3.

20 Essay: Examine the modifications that farmers have made to the physical environment in Tamil Nadu and eastern Tanzania. What do these suggest about the limits set by the physical resources?

Summary

- At the global scale, crop distributions owe most to precipitation and temperature.
- At the local scale, a farmer's choice of crops depends more on the soil texture and drainage of each field.
- In general, as scale gets smaller, climatic influences become weaker and soil quality and slopes become more important.
- Farmers have only limited ability to modify the physical environment to suit agriculture, but the right combination of physical factors at the local scale can give optimal conditions for crop growth.
- Modifications to climate and relief are especially limited. They include irrigation, glasshouse cultivation and terracing of hillslopes.
- Physical factors set limits on what farmers can grow, but apart from extreme environments they rarely determine what is grown.

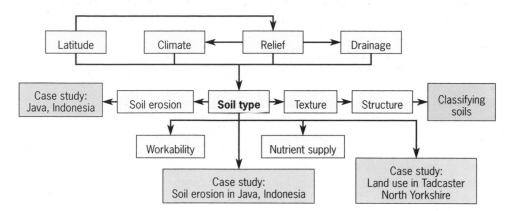

5 Soils and agriculture

5.1 Introduction

Soils are the mixture of mineral and **organic** matter in which plants grow, and are a major influence on agriculture. They supply crops with water for **photosynthesis** and **transpiration**, nutrients for growth, and a material in which to root.

5.2 The global scale

In the northern hemisphere, soils extend across Eurasia (Figs 5.1–5.5) and North America in a series of broad latitudinal belts. Such a distribution closely matches those of climate and vegetation, which, as we shall see in the next section, are mainly responsible for soil formation at this scale.

Low latitudes

Given the close relationship between climate and soils, it is not easy to isolate the influence of soils on agriculture. For example, in a harsh environment like a hot desert, soils are often saline (salty) and of little agricultural value. But while poor soils undoubtedly limit farming, other problems may be crucial. Not least of these is aridity, which imposes even greater restrictions on cultivation. In contrast, equatorial lowland climates, with high temperatures and high rainfall all year round, appear to favour farming. Yet surprisingly these areas have limited agricultural potential. This is because most tropical soils are infertile. They are strongly **leached** (see Section 4.3) and acidic, and contain few plant nutrients: after just a few months of cultivation they are quickly exhausted. In rainforest areas, local farmers have responded ingeniously to this challenge by developing systems of non-permanent, or **shifting cultivation** (see Section 3.6).

Middle latitudes

In middle latitudes, climate is less of an obstacle to agriculture. Here the differences in farming systems more often reflect differences in soil types. There are four major zonal soil types in middle latitudes: **chestnut soils**, **brown earths**, **chernozems**, and **podsols**. These soils, which owe their main characteristics to climate, provide varying opportunities to farmers, from **extensive** grazing for livestock to **intensive** arable cultivation (Figs 5.2–5.5).

1a Select four or five photographs from elsewhere in this book and locate these places on a world map.
b Match each place with its soil type, using evidence from the photograph you selected and the information here (Figs. 5.1–5.5).

Brown earths: associated with temperate deciduous forest. They are only slightly acidic; have immense argricultural potential and are extensively cultivated.

Podsols: found in higher latitudinal areas where precipitation exceeds evapotranspiration. Posols are acidic, sandy and have little agricultural value.

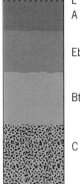

L	Acid mull humus
A	Mixed mineral and organic horizon, strongly acid
Eb	Lighter-coloured and eluvial (leached) strongly acid and depleted of clay
Bt	Illuvial horizon of deposition with clay enrichment moderately or strongly acid
C	Little altered parent material such as Keuper marl or boulder clay, possibly calcareous

**Figure 5.4
Profile of brown earth**

L F H	Mor or moder
A	Thin mixed mineral and organic horizon, strongly acid
Eb	Eluvial horizon bleached, depleted of iron, strongly acid
Bh Bfe	Black-coloured illuvial horizon enriched with organic matter with indurated horizon below enriched with iron
Bs	Orange-brown illuvial horizon enriched with iron, strongly acid
C	Parent material: little altered sands and gravels or sandstone

Figure 5.5 Profile of podsol

Soil types

	Tundra soils
	Podsol soils
	Brown earth soils
	Steppe soils
	Chernozem soils
	Chestnut soils
	Alpine soils
	Lateritic soils

Zonal soils are so-called because they correspond with broad latitudinal belts in Eurasia and North America. Soils owe their main characteristics to climate and the balance of precipitation and evaporation. However, natural vegetation cover has also been important in their development.

Figure 5.1 Zonal soils in Eurasia

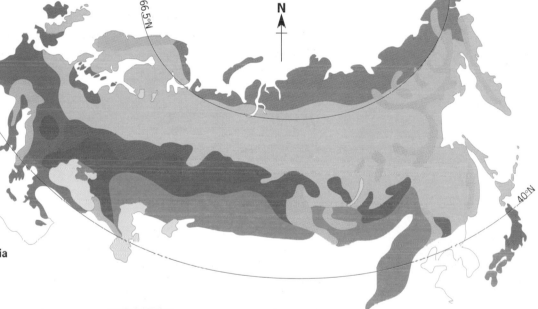

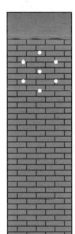

A	Thin horizon with limited supply of organic matter from grasses
Cca	Silty loam, highly calcareous. Accumulation of calcium carbonate, slightly alkaline.
C	Parent material in Eurasia mainly loess: in North America it may be calcareous glacial drift

Figure 5.2 Profile of chestnut soil

A	Mull humus incorporated to considerable depth by earthworms, neutral or slightly acid
	Krotovinas (burrows) of invertebrate animals
Cca	Parent material of loess or loess-like loams; concentration of calcium carbonate in Cca horizon but depth varies according to the amount of leaching
C	

Figure 5.3 Profile of chernozem soil

Chestnut (left) and chernozem soils: found in grasslands and continental interiors. They are neutral to alkaline, with a high humus content making them fertile. They are cultivated for cereal crops.

?

2 Study Figure 5.6 and Table 5.1. You are an adviser from the Ministry of Agriculture, Fisheries and Food. Assess the potential of agricultural land in areas **a** to **h**. Your results will be used as a basis for grants and compensation to farmers. Summarise your assessment as a table in which you justify your classification of each area.

a ⎱
b ⎰ Leached brown
c ⎱ soils
d ⎰

e ⎱ Acid brown
f ⎰ soils

g ⎱ Surface-water
h ⎰ gley soils

Figure 5.6 The soils of Wenlock Edge, Shropshire, UK (*Source:* Mackney and Burnham, 1965)

Arable land comprises classes 1 to 4, with classes 5 to 7 being more suitable to grassland, forestry and non-agricultural nature reserves. Each class is then divided into sub-classes according to limiting factors. Five limiting factors are recognised: wetness (w), soils (s), gradient (g), climate (c), and erosion (e). Although economic factors such as distance to market and the size of farms are not taken into account, the classification gives farmers an idea of the enterprises which are likely to maximise production.

5.3 The local scale

At a local scale, geographical differences in agricultural land use are closely related to soil texture (see Section 5.4), drainage, slopes and climate (see Sections 4.2–4.3). Based on these factors, farmland in the UK is graded by the Ministry of Agriculture, Fisheries and Food (MAFF) into seven categories according to its potential for cultivation (Table 5.1).

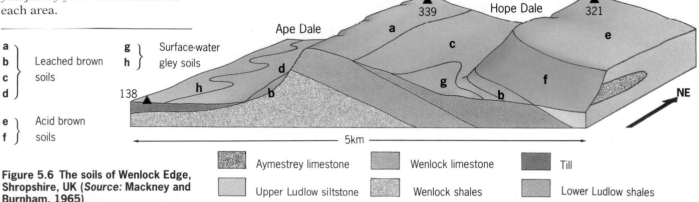

Table 5.1 Land capability classification of the UK (*Source:* MAFF)

Class 1 Land with very minor or no physical limitations to use.

Class 2 Land with minor limitations that reduce the choice of crops and interfere with cultivation.

Class 3 Land with moderate limitations that restrict the choice of crops and/or demand careful management.

Class 4 Land with moderate limitations that restrict the choice of crops and/or require very careful management.

Class 5 Land with severe limitations that restrict its use to pasture, forestry and recreation.

Class 6 Land with very severe limitations that restrict its use to rough grazing, forestry and recreation.

Class 7 Land with extremely severe limitations that cannot be rectified.

5.4 Physical characteristics of soils

Soil formation results from a complicated interaction between physical and biological processes. At the global scale, soils owe their main characteristics to climate and vegetation. Locally, rock type (known as the parent material), along with relief and drainage, have the most influence on soil formation (Fig. 5.7).

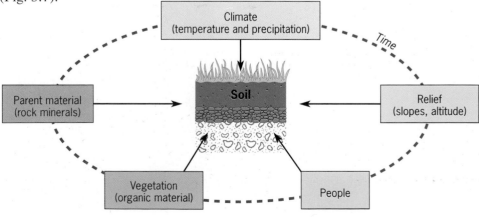

Figure 5.7 Soil-forming factors

Crop yields and the choice of crops available to farmers are strongly influenced by the physical and chemical characteristics of soils. However, farmers are able to modify soils and improve their potential for cultivation to a far greater extent than either climate or relief.

In our examination of soils we need to focus on two distinct properties: texture and structure. It is important to understand the differences.

Soil texture

Soil texture refers to the size of mineral particles in the soil. As far as farmers are concerned, it is the single most important soil characteristic.

?

3 Using Figures 5.8–5.11, identify the three types of soil (A, B, C) which have the following textural characteristics:

	% Sand	% Silt	% Clay
Soil A	60	30	10
Soil B	20	20	60
Soil C	40	40	20

Sandy soils: coarse textured and liable to summer drought. They have a limited store of nutrients but are light, easy to work and warm up quickly in spring. They have poor water-holding properties and crops on sandy soils easily suffer from drought.

Classifying soils

Soil texture is important because it controls soil drainage and a soil's ability to hold moisture. It also affects **field capacity** (see Section 4.4), workability and soil temperature. We group soils into textural classes according to their proportions of sand, silt and clay (Figs 5.8–5.11).

Figure 5.9 Profile of a clay soil

Clay soils: made up of tiny mineral particles. Clay soils are heavy and poorly drained. They are difficult to cultivate when wet or dry. They retain water easily, causing them to be frequently cold, heavy, poorly drained and difficult to plough. Waterlogging is common. They are slow to warm up in spring and are easily damaged by heavy machinery and trampling by livestock.

Loamy soils: the ideal soil texture, comprising a balanced mixture of sand, silt and clay. Loamy soils are highly valued for agriculture.

Figure 5.11 Profile of a sandy soil

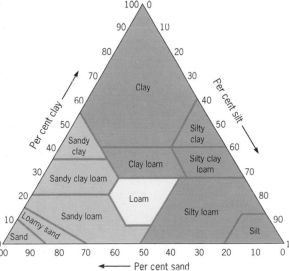

Figure 5.8 Classifying soil texture

Figure 5.10 Profile of a loamy soil

Classifying soils

4a Using Table 5.2, describe and explain how soil texture affects the return to field capacity.
b Explain the possible effect of this on cultivation.

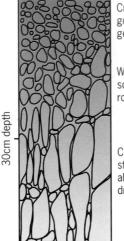

Crumb structure, good for seed germination

Well structured soil allowing root penetration

Columnar structures allowing good drainage

30cm depth

Figure 5.12 Well structured soil, valuable for agriculture

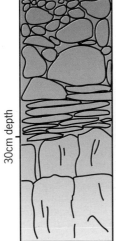

Blocky structures, making plant growth difficult

Platy structures, hindering root growth and water movement

Poorly drained subsoil

30cm depth

Figure 5.13 Poorly structured soil, of limited use for agriculture

Table 5.2 Soil texture and drainage

Location	Soil texture	Date of return to field capacity
Sandy, Beds.	Light	Late October–not at all
Spalding, Lincs.	Medium	Early September–early February
Leicestershire	Heavy	Early August–mid-January

Soil structure

Soil structure describes the way that individual mineral particles stick together to form larger units or **peds** (Figs 5.12–5.13). The effect of soil structure is to create voids, or gaps, which allow roots to penetrate and air and water to circulate, but soil structure is easily damaged in wet conditions by machinery and livestock. However, unlike texture, farmers can improve structure by adding organic material (or **humus**) to the soil.

5.5 Chemical characteristics of soils

Nutrient supply

Soils supply plants with the chemical elements (sixteen in all) which they need for growth and development. Of these elements, nitrogen (N), phosphorus (P) and potassium (K) are the most important. The ability of soils to retain these nutrients depends on soil texture and structure. Clay soils, for example, have a greater nutrient supply than sandy soils. An abundance of organic material or humus in a well-structured soil also increases its nutrient store.

In natural **ecosystems**, soil nutrients are continually re-cycled. However, this does not occur in **agro-ecosystems** because the harvesting of crops removes nutrients from the system. Farmers must therefore take steps to replace these nutrients if the soil is to retain its fertility (Table 5.3). Traditionally, they did this by spreading manure on the land. Today, in economically developed countries, farmers are increasingly using chemical fertilisers instead of organic ones. Unfortunately, the long-term use of such chemicals denies the soil essential humus, which is so important in preserving soil structure.

Table 5.3 Modifications to the soil by farmers

Problem	Response
Poor drainage	Underdraining of fields using clay pipes. Digging of dykes and drainage ditches.
Nutrient deficiency	Addition of organic (manure) or inorganic (N, P, K) fertilisers. Growing nitrogen-fixing crops, such as clover and vetch.
Moisture deficiency	Irrigation.
Acidity	Addition of lime.
Poor structure	Addition of organic material; sowing rotational grasses or **leys** (which have a dense root network).
Hard pan formation	Deep ploughing.

Land use in Tadcaster, North Yorkshire, England

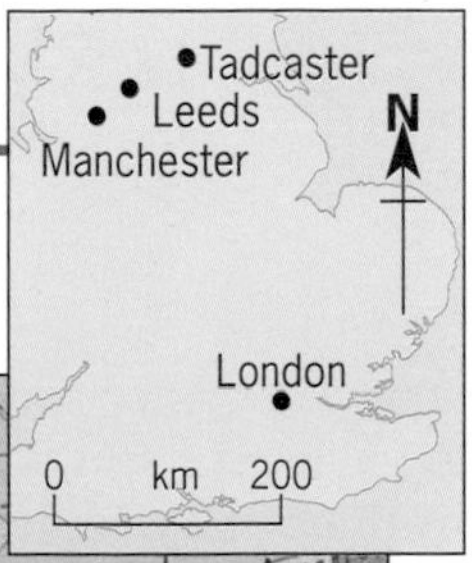

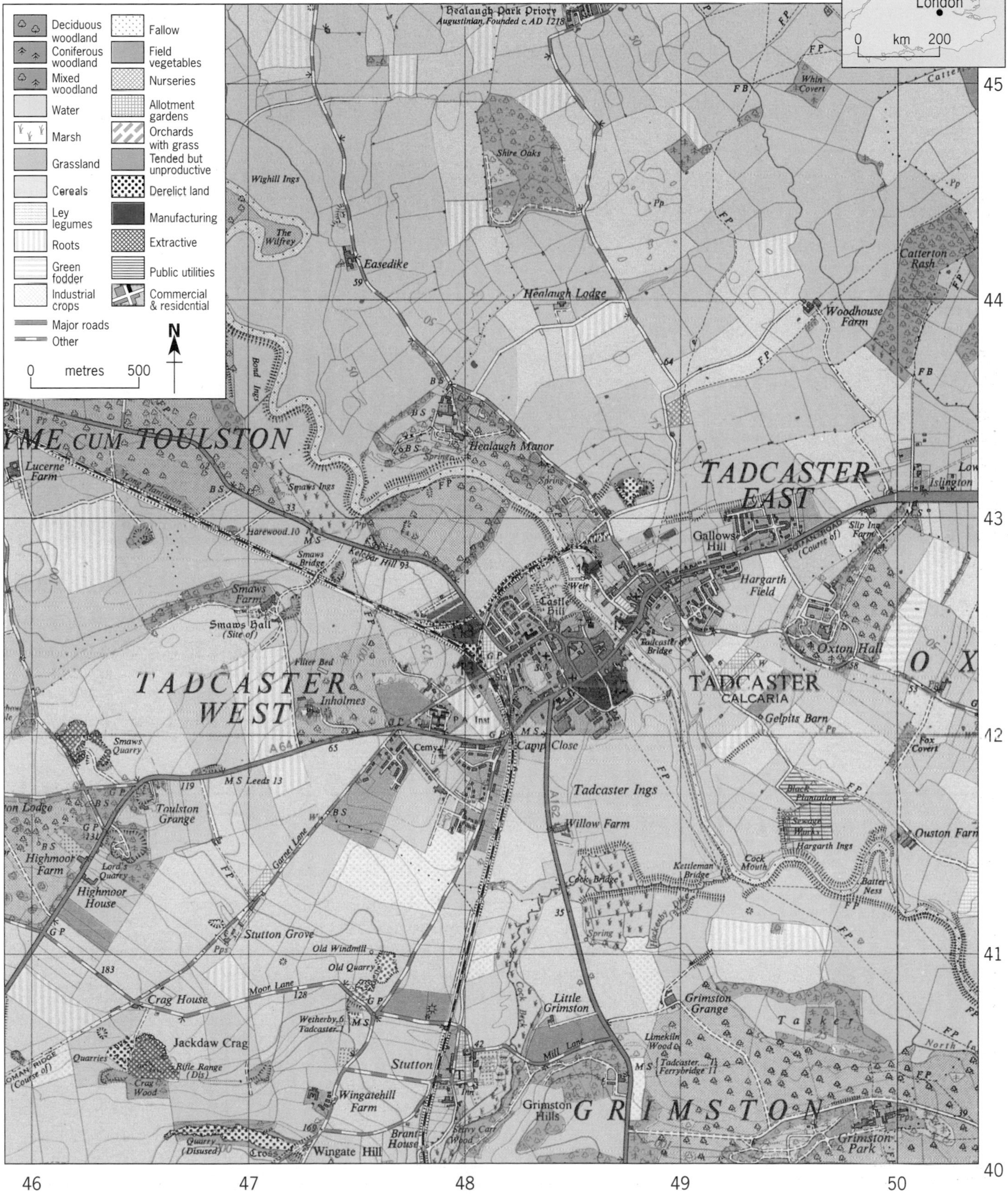

Figure 5.14 The Tadcaster area, North Yorkshire (© Crown Copyright)

Land use in Tadcaster

The Tadcaster area of North Yorkshire (Fig. 5.14) has a variety of soil types (Fig. 5.15). Does soil type influence agricultural land use in this area? Answer this question for yourself by completing the following exercises.

?

5a Using random numbers, generate 100 six-figure grid references which cover the study area. (If you work in small groups you can speed up this part of the exercise.)
b For each random grid reference, use Figures 5.14 and 5.15 to identify its agricultural land use and soil type. Use the following land use classes: arable, grassland, other agricultural.
c Make a contingency table (like Table 5.4) to structure your analysis.

Table 5.4 Classification of grid references by land use and soil types

Soil type	Land use		
	Arable	Grassland	Other
Loams			
Clays			
Alluvium			

d Calculate the chi-squared value to determine the statistical significance of the results in Table 5.4 (see Appendix 2).
e Describe and explain the relationship between agricultural land use and soil type.

6 From the understanding you have gained so far in Chapters 4 and 5, suggest other possible factors which might influence the pattern of land use around Tadcaster.

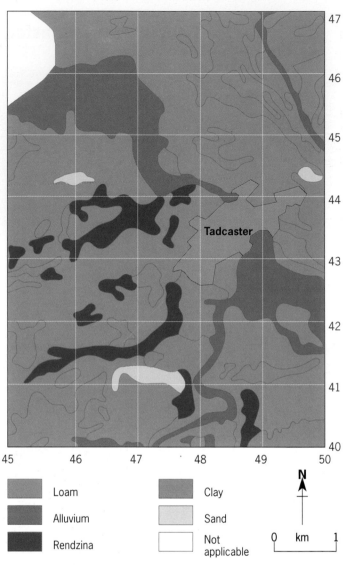

Figure 5.15 Soil types: Tadcaster, North Yorkshire

5.6 Soil erosion and human factors

Scientists have stressed the importance of physical factors like rainfall intensity and soil type in their explanations of soil erosion (see Section 6.4). However, we must not under-estimate the influence of human factors. In **subsistence agriculture**, the attitudes and traditions of farmers are particularly important. Also significant is rapid population growth which may lead to overcultivation and soil exhaustion. In **commercial agriculture** economic pressures contribute to soil erosion. This is because the demand for food and fibre cash crops encourages intensive farming which may ignore conservation practices, such as **contour ploughing**. Thus the search for short-term profit may lead to long-term ruin if the soil is eroded. The costs of such land degradation can, in fact, be calculated. If these costs were to be included in the price of crops, it might make both farmers and consumers think again.

Figure 5.16 Yatenga, Burkina Faso

7 Read the article in Figure 5.17.
a Construct a flow chart to illustrate the downward spiral of land degradation in Yatenga.
b In your opinion, what is the main cause of land degradation in Yatenga? Give reasons for your answer.

Creeping desert

In the beginning was a natural forest, too thick to walk through without a machete to hack your way. Then came shifting cultivators, cutting and burning the trees to make clearings for their crops, moving on to new ground after a year or two.

Every fifteen or twenty years they would come back to cultivate again. After a few of these cycles, the forest became an open woodland, with healthy stands of grass among the trees. Roots formed a dense mat through which rain percolated gently, soaking the soil below without eroding it.

From decade to decade the farmers returned to clear and burn the shrubs and trees that had regrown, and to grow crops. Just before the first rains each year, the soil was bare. The rains hammered down and some of the most fertile topsoil was washed away.

As the years progressed, and their numbers increased, the people returned more frequently. The soil became less fertile and the vegetation grew thinner. Each time the land was left to recover, the grass clusters were more widely spread. Foraging cattle and sheep chomped the grasses down to the base, leaving bare patches between the clumps.

Goats ate seedlings and saplings before they could mature. Wind and rain carried away more soil.

The Yatenga earth retaliated with a terrible vengeance. In the exposed patches, smaller soil particles washed off by the rain lodged between larger ones and formed an impervious crust. Rain could no longer seep down to plant roots, but ran off in sheets. In dry years and in wet years alike, there was now a drought in the soil.

Grasses were the first to die, surviving only temporarily in patches that the herds spared, or in slight hollows that collected soil stripped from the rest of the land. A year or two later, bushes withered and dried to rattling bundles of sticks.

Trees with their deeper roots hung on till the last, but finally succumbed as the water table dropped, no longer replenished each year by the rains. Their dead branches were cut for fuel, fences or timber and carried away. Later the weak roots were torn out.

What once was forest is now bare crust covered with gravel too heavy for the rain to wash way, a desiccated, desert land, useless to man or beast, where the rains scour destructively off.

Figure 5.17 A fragile future (Source: © Paul Harrison, Oxfam in association with the Observer, 1988)

Figure 5.18 Collecting firewood, Burkina Faso

Soil erosion in the UK

Soil erosion is a widespread and growing problem in the UK. In the intensive cropping areas of eastern England, soil losses typically reach 20 tonnes per hectare per year – double the global average. Erosion is caused by both runoff and wind, as well as other factors which are shown in Figure 5.19. Water erosion leads to the formation of rills and gullies (Fig. 5.20) and is more widespread than wind erosion. Although less significant, wind erosion is more spectacular (Fig. 5.21). The worst affected areas are the light soils of East Anglia, the Vale of York, Nottinghamshire and eastern Scotland.

8 You are a government adviser employed by the MAFF. You are asked to speak at a farmers' meeting in East Anglia about the causes of soil erosion. Write a speech explaining:
a the practices which can lead to soil erosion.
b possible measures which might help to reduce soil losses.

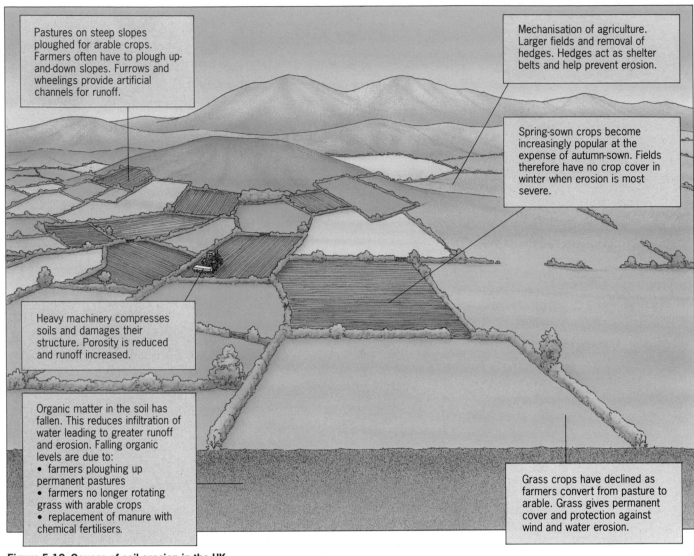

Pastures on steep slopes ploughed for arable crops. Farmers often have to plough up-and-down slopes. Furrows and wheelings provide artificial channels for runoff.

Mechanisation of agriculture. Larger fields and removal of hedges. Hedges act as shelter belts and help prevent erosion.

Spring-sown crops become increasingly popular at the expense of autumn-sown. Fields therefore have no crop cover in winter when erosion is most severe.

Heavy machinery compresses soils and damages their structure. Porosity is reduced and runoff increased.

Organic matter in the soil has fallen. This reduces infiltration of water leading to greater runoff and erosion. Falling organic levels are due to:
• farmers ploughing up permanent pastures
• farmers no longer rotating grass with arable crops
• replacement of manure with chemical fertilisers.

Grass crops have declined as farmers convert from pasture to arable. Grass gives permanent cover and protection against wind and water erosion.

Figure 5.19 Causes of soil erosion in the UK

Figure 5.20 Soil erosion caused by runoff

Figure 5.21 Soil erosion caused by wind

Soil erosion in Java, Indonesia

The Indonesian island of Java (Fig. 5.22) has very high rural population densities (more than 500 per square kilometre). Intensive cultivation of tiny plots is therefore needed to support the bulk of the rural population (Fig. 5.23). As the population pressure is so great, half of all holdings are less than one hectare in size, and one-third of the rural population is landless. The result of such pressure is a high rate of soil erosion (Table 5.5), especially on limestone/marl soils, where losses are between 19 and 60 tonnes per hectare per year. The most damaging land use is *tegal* or rain-fed cropping of maize, rice and cassava on sloping upland fields.

The costs of soil erosion are difficult to assess. However, best estimates put siltation damage to **irrigation** systems, reservoirs and harbours at £20–60 million per year, and losses due to reduced yields at £230 million per year (or four per cent of the value of *tegal* crops).

Government policies, assisted by overseas aid, have aimed at reducing soil erosion. In fact, the Indonesian government has even tried to reduce population

Figure 5.23 Javanese 'household gardens': these are complex food-producing ecosystems carefully managed and organised to maximise output

Table 5.5 Indonesian land use and soil erosion, 1990

	Area (million hectares)	Erosion (tonnes/hectare/year)
Sawah (wet rice)	4.6	0.5
Forest	2.4	5.8
Degraded forest	0.4	87.2
Wetlands	0.1	0
Tegal	5.3	138.3

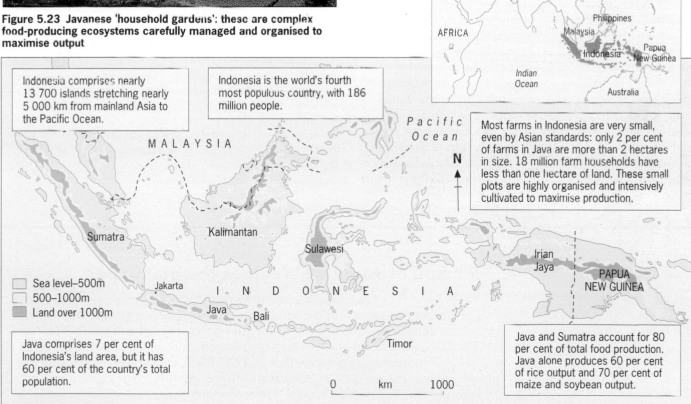

Indonesia comprises nearly 13 700 islands stretching nearly 5 000 km from mainland Asia to the Pacific Ocean.

Indonesia is the world's fourth most populous country, with 186 million people.

Most farms in Indonesia are very small, even by Asian standards: only 2 per cent of farms in Java are more than 2 hectares in size. 18 million farm households have less than one hectare of land. These small plots are highly organised and intensively cultivated to maximise production.

Java comprises 7 per cent of Indonesia's land area, but it has 60 per cent of the country's total population.

Java and Sumatra account for 80 per cent of total food production. Java alone produces 60 per cent of rice output and 70 per cent of maize and soybean output.

Sea level–500m
500–1000m
Land over 1000m

Figure 5.22 Indonesia

Soil erosion in Java

pressure by encouraging transmigration from Java to less densely populated outer islands. It also gives advice to farmers on soil conservation measures.

Soil erosion: economic, social and political causes

Links to the world economy

In the early 1980s, Indonesia faced a series of severe economic shocks. The most critical was the 60 per cent drop in oil prices between 1982 and 1986. Oil production was the backbone of the economy, providing 80 per cent of export earnings. Worldwide recession led to a reduction in demand for Indonesia's exports, while the depreciation in the US dollar after 1985 meant that world commodity prices fell. Overall, the effect was to weaken the Indonesian economy, with the government having to borrow heavily.

Cash crops

These economic pressures forced the government to promote the export of cash crops, such as cassava and vegetables, to pay back loans. (The decision of the European Community to import ten per cent of Indonesia's cassava crop gave an additional boost to this policy.) A substantial increase in the price paid to farmers for cassava brought an overwhelming response. It caused many farmers to abandon their traditional mixed farming systems and switch to a **monoculture** of cassava. In some cases farmers have even removed terraces to increase the cropped area. Meanwhile, vegetable growing has become highly profitable. Farmers have responded by intensifying cultivation, often on steeply-sloping volcanic soils which, though fertile, are easily eroded. Vegetables and sugar cane are usually grown by **share tenants** on land owned by absentee landlords. Share tenants, who do not own their farms and whose rent is a fixed

percentage of their harvest, have little incentive to conserve the soil. Thus they often resort to exploitative farming methods which make erosion even worse.

Results

Inevitably, soil erosion and other natural resource degradation have occurred on a massive scale, especially in Java. Soil erosion alone costs an estimated US$400 million a year in Java. Rivers, streams and irrigation canals are silted; low-lying areas are flooded; coastal fisheries are damaged; and fertiliser and pesticide runoff is polluting water supplies.

This results in considerable difficulties for the Indonesian government. There are so many farming types in the country that a single blanket policy for agriculture and food has little value. Lowland farmers have been subsidised with fertilisers and pesticides in order to boost national self-sufficiency in basic foodstuffs. However, this has led to the neglect of rain-fed upland farming. More research into problems of farming on erodible soils might help policy-making for sustainable development in the long-term – something that artificial subsidies for cassava and vegetables will not.

9 You are an official of the World Bank, which gives loans to economically developing countries to assist development. Your job is to advise on giving further loans to Indonesia to help it overcome its economic crisis. Any new loan is conditional on the Indonesian government modifying its agricultural policies. Write a report for the World Bank stating in your view:
a what is wrong with the current policy
b how the policy should be changed.
c Give reasons for your views.

5.7 Soil erosion: where does the blame lie?

Table 5.6 sets out two extreme perceptions of the soil erosion problem. Both the environmentalist and the economist have the same attitude: they regard soil erosion as a serious problem and one which needs immediate attention. However, their beliefs cause them to disagree on the causes of the problem. The environmentalist believes that soil erosion is caused by the structure of society and the unequal distribution of power within it; the economist believes that soil erosion results from mismanagement of environmental resources and that farmers are largely to blame.

10a Consider the views presented in Table 5.6.
b Assess your own attitude to soil erosion by noting the perceptions in Table 5.6 with which you agree.
c Present your views as a journalist writing an article for a magazine such as *The Economist* or *New Scientist*.

11 Essay: With reference to an economically developed country and an economically developing country, examine:
a the causes of soil erosion,
b the methods which have been used to solve the problem of soil erosion.

Table 5.6 Perceptions of the soil erosion problem in the developing world

Values motivated by the structure of society and the distribution of power ←--- { Political spectrum } ---→ Values motivated by the management of resources and the production and consumption of goods

Values motivated by the structure of society and the distribution of power	Values motivated by the management of resources and the production and consumption of goods
• Soil erosion is a sympton of an unjust society. Rich people own most of the land, which forces poorer people to farm intensively on the remaining land. They cannot afford to pay for anti-erosion measures.	• Soil erosion is largely caused by poor farming methods and over-population. Too many people create a demand for food which the available land cannot supply on a sustainable basis. So, measures must be taken to help farmers and reduce population growth.
• Economic development is in the interests of wealthy urban élites. They are powerful and have little concern for peasant farmers and land degradation. Thus resources, including aid, are not directed to rural areas. When they are, they most often reach larger commercial farmers whose attitude to farming is often exploitative and damaging.	• People need laws and financial penalties to encourage them to combat soil erosion. Grazing should be banned in vulnerable areas; people should be required to build terraces and other conservation works; soil conservation education should be introduced to counter ignorance and demonstrate new techniques. Aid must be tied to making farmers conserve the soil and adopt technologies which allow greater, sustainable food production.
• Soil erosion will only be tackled by taking land away from wealthy landowners and sharing it out fairly.	• Soil erosion will only be tackled by making the farmers follow strict directives and encouraging them in conservation.

Summary

- Soils are the mixture of mineral and organic matter in which plants grow.
- Globally, soil distributions closely match climate and vegetation belts.
- Locally, soil type depends on parent material, relief and drainage.
- Agricultural land use is determined by the physical and chemical characteristics of soils.
- Soils are the physical resource which offers greatest scope for modification through drainage, applying manure or adding chemical fertiliser.
- Farming activities, particularly in agro-ecosystems, often damage the quality of soils.
- Soil erosion is caused by both physical and human factors.
- Physical causes of soil erosion include wind and water.
- Human causes of soil erosion include the world economy, government incentives and poverty.
- Environmentalists and economists have different views on the causes of soil erosion.

6 Agricultural impacts on the environment

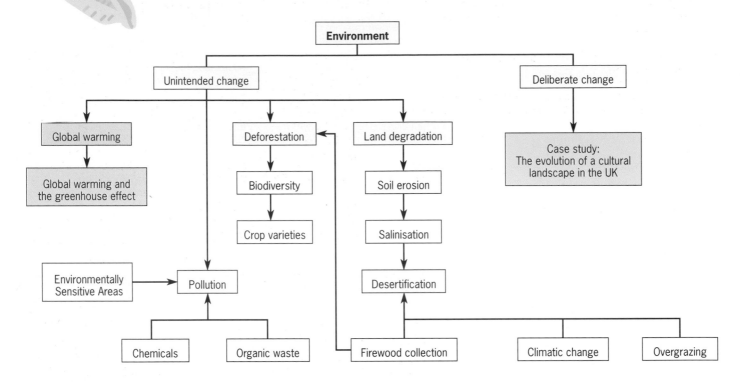

- Environment
  - Unintended change
    - Global warming
      - Global warming and the greenhouse effect
    - Deforestation
      - Biodiversity
        - Crop varieties
    - Land degradation
      - Soil erosion
        - Salinisation
          - Desertification
  - Deliberate change
    - Case study: The evolution of a cultural landscape in the UK

- Environmentally Sensitive Areas → Pollution
- Pollution ← Chemicals, Organic waste
- Firewood collection
- Climatic change
- Overgrazing

6.1 Introduction

There is no doubt that the physical environment has an enormous influence on agriculture (Chapters 4 and 5). However, agriculture has equally important effects on the physical environment. These effects may be deliberate or unintended. For instance, woodland clearance to make land available for cultivation is clearly deliberate. On the other hand, groundwater pollution by nitrate fertilisers is quite unintentional.

Agriculture has touched all aspects of the environment. Indeed, its influence is so widespread that it has created its own distinctive landscapes. For the most part these **cultural landscapes** evolved over hundreds of years. However, when people changed the landscape, they often did this suddenly: long periods of stability would be interrupted by sudden upheaval. This happened in much of lowland Britain in the eighteenth century when enclosure divided up the medieval open-fields. In this chapter we shall examine such changes; their complex economic, social and political causes; and their environmental impact.

6.2 Deliberate change: agriculture and the cultural landscape

Pre-industrial societies

Pre-industrial societies depend for their living either on **hunting** and **gathering** or **subsistence agriculture** (see Sections 3.6 and 2.3 respectively). Until recently the numbers of hunter-gatherers and subsistence farmers were small. This, together with their relatively simple technology, meant that they

Figure 6.1 Fourteenth century plough and team

?

1 Study the three landscapes in Figures 6.2–6.4, and the front cover. For each one describe how people have changed the landscape for farming.

had only a limited impact on the landscape. For pre-industrial societies, the most powerful 'tool' for landscape change was fire. However, as population grew and technology advanced, people's control over the environment increased, and with it their capacity to change the landscape. The agricultural activities listed in Table 6.1 were especially important to landscape change in the pre-industrial era.

Table 6.1 Modification of the agricultural landscape (*After*: Simmons, 1987)

- The use of fire to clear forests and grasslands for cultivation.
- The development of stone, and later metal axes speeded up woodland clearance.
- Modifications to the soil by the use of digging sticks, spades and ploughs.
- The building of terraces, mounds and ridge-and-furrow for agricultural use.
- Irrigation and artificial drainage works.
- The use of fences, dykes, ditches and bunds (embankments) as boundaries and livestock barriers.
- The domestication and selective breeding and spread of useful animals and plants.

Figure 6.3 A polder (reclaimed land) with dykes, Netherlands

Figure 6.2 Hoeing for potatoes, Bolivia

Figure 6.4 Clearing tropical rainforest, Indonesia

The evolution of a cultural landscape in the UK

The rural landscape of the UK has been moulded by the agriculture and technology of many different cultural groups over the last 5000 years. Today's cultural landscape therefore includes many relict features such as fields corrugated by **ridge-and-furrow**, and hedges. Although these are no longer useful to modern farmers, they are significant features of the cultural landscape and often appeal to us for sentimental and aesthetic reasons.

Ridge-and-furrow
Fields with wave-like patterns of ridges and furrows are a common sight in many parts of lowland

England (Figs 6.5 and 6.6).

Evidence of medieval ploughing
They owe their appearance to ploughing with heavy, ox-drawn ploughs from medieval times until the eighteenth century (Fig. 6.1). The reason for a ridge-and-furrow was to improve land drainage: by ploughing downslope, water ran off the fields, with the furrows acting as drainage channels. Ridge-and-furrow also helped peasant farmers to identify easily their own strips of land in the medieval open-field system.

Cultural landscape in UK

Figure 6.5 The village of Padbury in Buckinghamshire, surrounded by fields corrugated by medieval ridge-and-furrow

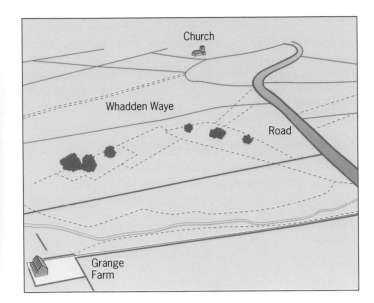

Figure 6.6 Field patterns around Padbury

Technological response

By the eighteenth century the ridge-and-furrow technique fell out of use. Instead, farmers were now draining fields using clay pipes, while open-field agriculture had long since been abandoned. Despite these changes, ridge-and-furrow is still visible today: you can see evidence of it where farmers have grassed over arable fields because the price of livestock products made sheep and cattle grazing more profitable than cereal cultivation. Signs of the ridges and furrows also remain inspite of their inconvenience for mechanised farming. Some high-back ridges on heavy clay soils rise as much as 1.5 metres above the level of adjacent furrows. Ploughing out the ridges is simply too expensive, and would only result in exposing less fertile sub-soil.

Hedges in the agricultural lowlands

Hedge boundaries, enclosing a patchwork of small fields, are a characteristic feature of the UK's cultural landscape. Indeed, to many people they are the essence of the British countryside (Fig. 6.5) and, like ridge-and-furrow, they are a relic of past farming practices. They are also inconvenient to modern arable farming. Farmers can only operate large machines, such as combine harvesters, efficiently

?

2 Study Figures 6.5 and 6.6
a Describe and explain the alignment of strips in relation to the stream.
b Make two copies of Figure 6.6 to show field patterns and settlements before and after enclosure. Add annotations to your sketches to bring out the main elements of landscape change.

where the fields are themselves large. For example, in cultivating a 2 ha field with a wide machine, farmers can waste two-thirds of their time turning and dealing with difficult corners. By comparison, in a 40 ha field this wasted time is reduced to 20 per cent.

Disappearing hedges

As a result, farmers have a strong incentive to increase field sizes. They have responded by grubbing-up hedges and removing hedgerow trees at an alarming rate. Between 1947 and 1990, 377 000 km of hedges (nearly half the UK total) were destroyed. The loss of habitat for wildlife, especially birds, and the effects on hedgerow plants, have been devastating. None the less, for many years the government subsidised farmers to remove hedges and it was only after strong opposition from conservationists that this policy was revised. Since 1992, new policies have encouraged traditional management such as hedge-laying and coppicing.

Enclosure in the agricultural uplands

Large areas of moorland at an altitude of over 250 m were enclosed in the central Pennines in Lancashire between 1770 and 1840 (Figs 6.7 and 6.8). The aim of enclosure was to improve land which, 'from generation to generation had remained underdeveloped and often enormously overstocked' (Harrison, 1888). It meant that people could own land privately, whereas it had previously been owned in common.

With privatisation, people limed and drained the moors, and even brought small areas into cultivation. Even though people's environmental resources for farming were limited, they grew oats on the better land and reclaimed some meadow. However, most land was left as low-grade permanent pasture.

Enclosure also caused social change, so that only the common rights of the lord of the manor and freeholders were respected. Consequently, many small farmers who had depended on common grazing rights to make a living now found that their holdings were too small to be viable. They also lost access to other resources such as free fuel, and the right to collect briars and gorse. Meanwhile squatters, who previously had been allowed to build cottages on the commons, were evicted.

3a Look at Figures 6.7 and 6.8. Copy one of the maps and use it as an overlay to the other map. On it, highlight the changes in field patterns which occurred after enclosure.
b Describe how the landscape of Oswaldtwistle Moor changed between 1776 and 1845. Make particular reference to: field boundaries; settlement pattern; density of settlement; and road pattern.
c Draw an annotated diagram to show the forces responsible for this change.

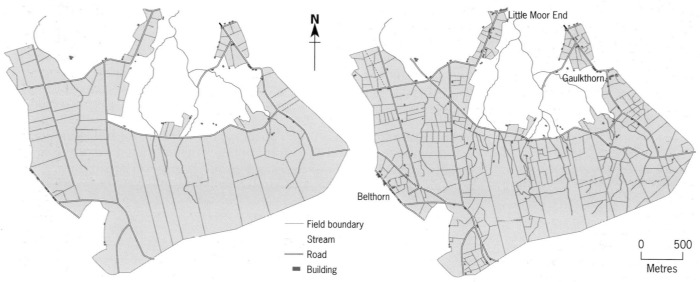

Figure 6.7 Oswaldtwistle Moor, Lancashire, 1776: before enclosure

Figure 6.8 Oswaldtwistle Moor, Lancashire, 1845: after enclosure

6.3 Unintended change: agriculture and land degradation

4 Study the information in Figure 6.9.
a Which continent has the largest area of land classed as dryland?
b Which continent has the largest: • area, • percentage, of degraded dryland?
c Describe the overall differences in percentage land degradation between drylands and other lands. Suggest reasons for these differences.

Agriculture makes cultural landscapes by changing existing landscape forms and creating new ones. Unfortunately, the impact of agriculture has often been destructive. At worst, **intensive cultivation** and the use of chemicals have resulted in **land degradation** which, if left unchecked, has serious consequences both for food production and the physical environment. Land degradation describes the end result of a range of processes which include soil erosion, **salinisation** and **desertification**. Drylands are especially vulnerable to these processes (Fig. 6.9).

Figure 6.9 Land degradation by continent

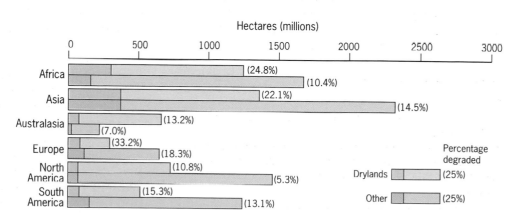

6.4 Unintended change: soil erosion

It is important to understand at the outset that soil erosion occurs naturally. Natural grassland or woodland loses about 0.1 tonnes of soil per hectare per year through erosion. However, this loss is balanced by **inputs** of new soil by the weathering of rocks and the decay of plants. Thus the effect of human activities (especially agriculture) is not to cause soil erosion but greatly to accelerate it.

One measure of soil erosion is the volume of sediment currently transported by the world's rivers. Estimates suggest that today's sediment load is 2.7 to 5.0 times greater than before human disturbance of the landscape. At the end of the twentieth century soil erosion has become a global problem: modern agriculture is destroying its most fundamental resource at an alarming rate.

Soil erosion usually follows the destruction of the vegetation cover which protects the topsoil. For example, forest clearance may trigger erosion, especially on soils exposed to heavy tropical downpours. Elsewhere, overgrazing or lack of soil conservation on slopes exposes soils to erosion by wind or water. Soil type is also crucial when considering erosion. The easily erodible **loess** soils of the Yellow River basin in China lose 100 tonnes per hectare annually (Fig. 6.10). This is ten times higher than the world average! Globally, around 26 billion tonnes of topsoil are lost every year.

Figure 6.10 Eroded loess soil, Shaanxi Province, China

The effects of soil erosion on farming

Soil erosion (see Sections 5.6–5.7) has three major effects on the environment. It causes a loss of soil nutrients and **organic matter** which reduces the natural fertility and water-retaining capacity of the soil (see Figure 5.16). It may carve deep channels or gullies across fields, which make farming difficult. And it increases sediment loads in streams and rivers, causing problems of silting in **irrigation** canals, dams and reservoirs. Siltation also increases the risk of flooding downstream.

6.5 Unintended change: salinisation

Salinisation is a form of land degradation found mainly in arid and semi-arid climates, affecting seven per cent of the world's land area. It describes the accumulation of chloride, sulphate and carbonate salts of sodium, calcium and magnesium in the soil (Fig. 6.11). This occurs when **evaporation** and

?

5 Use Figure 6.11. Write a personal response to the extent of the problem shown here.

Figure 6.11 Salinated soil, Sind Province, Pakistan

transpiration are greater than **precipitation** in areas where the water table is near to the surface. The presence of salts in soils seriously affects crop yields, because salts reduce the soil's capacity to hold air and nutrients. They are also toxic to many plants.

Figure 6.12 The process of salinisation in irrigated areas in the tropics and sub-tropics

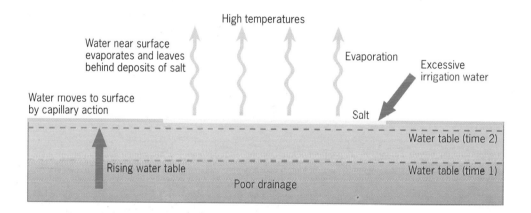

High temperatures

Water near surface evaporates and leaves behind deposits of salt

Evaporation

Excessive irrigation water

Water moves to surface by capillary action

Salt

Water table (time 2)

Rising water table

Water table (time 1)

Poor drainage

Like soil erosion, salinisation occurs naturally, but is often made worse by human activity. For example, the risk of salinisation is particularly high when farmers practise irrigation agriculture without adequate drainage. This causes a rise in the water table. Salty groundwater may reach the root zone of crops, or even worse, be drawn to the surface by **capillary action** (Fig. 6.12). In Australia, over 40 per cent of soils are affected by salinisation, with the worst problems in the irrigated areas of South Australia and Victoria. In Syria 50 per cent of irrigated land is salinised, and in Uzbekistan up to 80 per cent of some large-scale irrigation projects have been lost to this process.

Dealing with the problem of salt accumulation is expensive. Drainage systems to lower water tables and dispose of salty water can be installed at a cost, and in recent years the World Bank has financed successful drainage schemes in the Nile Valley in Egypt. A cheaper remedy, though, is to change land use from crops to pasture, or to switch to more salt-tolerant crops such as barley, sugar beet and cotton.

?

6a Use Figure 6.12 to explain the changes in the soil system which cause plants to die.
b In what ways would you choose to deal with salinisation? Explain your answer.

7 Study Figure 6.13.

a Name a country from each continent which has severe desertification.

b Which continent suffers most from: • moderate desertification, • desertification? Explain your choice.

6.6 Unintended change: desertification

Desertification describes the human and climatic processes which reduce biological activity to a point where desert-like conditions prevail. The popular image of desertification is one of irreversible change, with sand dunes encroaching on over-grazed, semi-arid pastures. The reality is more complex. Desertification is found in a wide range of environments (Fig. 6.13). In 1991, it affected 850 million people – a figure which will probably rise to 1.2 billion by the year 2000. Each year desertification results in crop losses of £28 billion and causes 6 million hectares of farmland to go out of production.

Causes of desertification

Desertification occurs in a number of different environments. This suggests that it is a complex process with many different causes.

Watering holes

At a local scale, desertification is common around water holes in desert fringe areas (Fig. 6.14). Here, large numbers of livestock gather, and this leads to overgrazing and degradation of the vegetation. The problem is especially acute in drought years when **nomadic pastoralists** may take up semi-permanent residence around water holes.

Firewood

In economically developing countries the demand for firewood for cooking and lighting is a major cause of desertification (see Fig. 5.18). Many African countries, such as Mali, Burkina Faso and Tanzania, depend on firewood for over 80 per cent of their domestic energy. In densely populated rural areas,

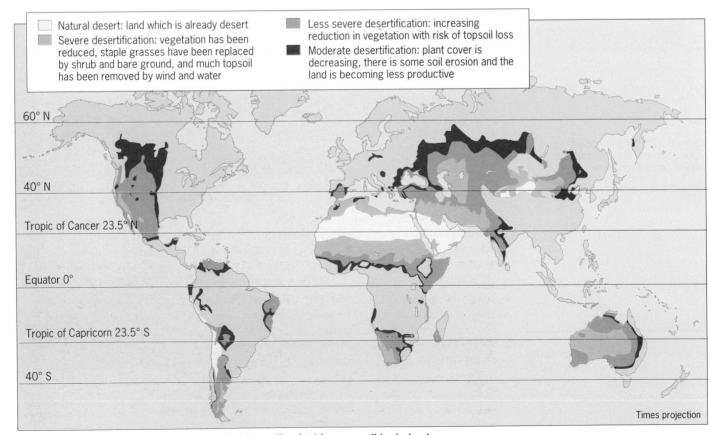

Figure 6.13 Global soil degradation severity (desertification) in susceptible drylands

8a In Figure 6.13, between which latitudes is severe desertification found?
b Give reasons which contribute to this distribution.

9 Comment on the following hypothesis: desertification is a global rather than a developing-world problem.

Figure 6.14 Livestock waiting to be watered, Senegal

the demand for firewood may result in people clearing all trees and shrubs within a zone of two or three kilometres around their village. The subsequent loss of the binding effect of vegetation makes the soils in these areas vulnerable to erosion by both wind and water.

Climate, soil and cultivation

Desertification is also closely linked with climatic change (especially drought) and soil type. Unfavourable climatic cycles such as the 1968–73 drought in the Sahel on the southern edge of the Sahara, and the dry period in southern Africa in the 1980s, caused widespread desertification in both locations. Finally, over-cultivation of soils which have limited fertility and fragile structures may lead to desertification and land abandonment.

6.7 Unintended change: deforestation

Early farmers, from stone-age to medieval times, cleared woodlands by burning, felling and animal grazing. Between 1500 and 1700, most of the temperate deciduous forests were destroyed across Europe.

During the 1980s, 152 million hectares of tropical forest were cleared (an area more than six times the size of the UK). Asia has already lost 42 per cent of its rainforest; Africa and Latin America have each lost 37 per cent.

Although commercial logging and informal cutting of fuelwood are important causes of deforestation, the major factor is clearance for agriculture (Fig. 6.3). Central America has already lost two-thirds of its forests to plantations and ranching for hamburger beef.

Unlike traditional rainforest agriculture, such as **shifting cultivation**, recent farming developments in the rainforests are unsustainable. This is because conventional harvesting following rainforest removal quickly damages the fragile natural ecosystem. New roads constructed by governments and commercial companies have opened up vast areas to large-scale ranching, plantations and peasant farming. The result is that after only a few years of cultivation soils are exhausted, the environment is degraded, and the land is often abandoned. Forest clearance is so thorough that regeneration is out of the question.

?

10 Test the hypothesis that deforestation is linked with rural population pressure in Africa by using the data in Table 6.2 and completing the following tasks:
a Plot the annual rate of deforestation (*y*) against the population density per km² of farm land (*x*) as a scattergraph.
b Describe the relationship between the variables *x* and *y*.
c Calculate the Spearman rank correlation coefficient between *x* and *y*, and its significance level (see Appendix 1). Comment on the validity of the hypothesis.
d Explain how deforestation might lead to further environmental degradation in Africa.

Table 6.2 Deforestation and population pressure in Africa

	Annual rates of deforestation (%)	Population density per km² of farm land (1993)
Angola	−0.2	40.8
Cameroon	−0.4	109.2
Ethiopia	−0.1	112.4
Ghana	−0.8	328.7
Kenya	−0.8	569.5
Mozambique	−0.8	43.5
Nigeria	−1.9	286.0
Senegal	0	79.7
Sierra Leone	−0.2	135.9
Tanzania	−0.3	98.6
Uganda	−0.8	229.6

6.8 Deliberate and unintended change: reduction of biodiversity

In forests
Almost without exception, agriculture simplifies natural ecosystems and reduces their species richness or **biodiversity**. An example of this is tropical forests which have a remarkable biological diversity because they contain nearly two-thirds of all plant species. Forest clearance threatens this diversity and could mean the loss of one quarter of the world's species before the end of the twentieth century. Such destruction is both wasteful and foolish. The forests contain a vast genetic pool with enormous value for use in medicine and agriculture. Once this resource is lost, it cannot be replaced.

In agriculture
Modern agriculture focuses production on a small number of crop varieties, thus further reducing biodiversity (Table 6.3). Farmers have abandoned many traditional varieties either because their yields are too low or because they are non-marketable. For example, of the 2000 apple varieties grown in France one hundred years ago, only about a dozen remain (Fig. 6.15). Similarly, in Greece, 95 per cent of native wheat varieties ceased commercial production between 1945 and 1986.

Results
This loss of genetic diversity is dangerous. Unlike many new high-yielding varieties (HYVs) of cereals recently produced by plant breeders (see Section 11.4), traditional varieties are often resistant to pests and diseases. Furthermore, the HYVs have little value in marginal areas where environmental conditions are far from optimal (see Section 4.2) for cultivation . It has not been possible to combine all the desirable genes in a single variety.

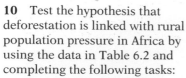

Figure 6.15 Picking apples for calvados and cider, France

Table 6.3 The loss of vegetable varieties

Vegetable	Number of varieties available in 1903	Number of these left in 1992
Artichoke	34	2
Asparagus	46	1
Runner bean	14	1
Lima bean	96	8
Garden bean	578	32
Beetroot	288	17

11a Study Table 6.3. Use a suitable graph to represent these statistics.
b Suggest why fewer varieties of fruit and vegetables are available in the UK now than a hundred years ago.
c What is your own view on this reduction in diversity?

Biological diversity is invaluable and we simply cannot afford to lose the genetic reservoir of natural plant species and crop varieties. This reservoir almost certainly contains many plants with desirable characteristics such as drought resistance, disease resistance and salt tolerance that could benefit agriculture enormously in future.

6.9 Unintended change: agricultural pollution

For a more detailed study of chemical pollution, the use of fertilisers and the impact of nitrates on the environment, you should refer to Chaper 3. Here, we are only considering the changes caused by such products.

Toxins
There is widespread concern about the use of toxic chemicals in agriculture. Some of these chemicals, which are used in fertilisers, pesticides, herbicides and fungicides, are harmful not only to weeds and pests, but to other plants and animals, and also to people. Particularly harmful are those chemicals which collect in the soil and get into the food chain.

Natural waste
Even natural animal waste, which is normally welcomed in low-intensity organic farming, becomes a nuisance when too much is produced. In the Netherlands, there is a surplus of manure from dairy farming and intensive livestock rearing. Ninety-four million tonnes are produced each year, of which only 50 million tonnes can be absorbed safely by the land as fertiliser. Strict regulations concerning its use are enforced because nitrates and phosphates leached from manure can contaminate water supplies. Meanwhile in the UK, river pollution incidents, caused by accidental discharges of slurry tanks and silage stores, have increased alarmingly. In fact, agriculture is a major polluter of rivers throughout the economically developed world.

6.10 Unintended change: global warming

The earth is itself an ecosystem powered by solar energy. So long as there is a balance between energy input and **output**, the global ecosystem will remain stable. Indeed, except for minor fluctuations such as the glacials of the last two million years, the earth's climate has remained constant for long periods of time.

Global warming and the greenhouse effect

Until 1992, the eight warmest years of the twentieth century were between 1983 and 1992, and 1990 was the hottest year ever recorded. Is this a long-term trend, a minor fluctuation, or just coincidence? Scientists are not sure. But if the theory of **global warming** is correct, it will have major implications for world food production (Fig. 6.16).

Greenhouse effect
Although there may be uncertainty about the reality of global warming, there is general agreement about how it might be caused. Burning fossil fuels and forests releases carbon dioxide and other gases into the atmosphere.

Global warming and the greenhouse effect

Together with releases of methane, chlorofluorocarbons (CFCs) and nitrous oxides, these gases have a **greenhouse effect** (Table 6.4). This is because they are transparent to solar radiation but are very effective absorbers of the long-wave radiation emitted from the earth (Fig. 6.17). Thus, the more carbon dioxide there is in the atmosphere, the more heat is trapped from the earth's surface. In systems terms, energy output from the earth's atmosphere has fallen and no longer balances with the energy input. The result should be rising global temperatures and a slight rise in global mean precipitation.

Agriculture is affected in two ways: directly by the increased amounts of carbon dioxide available for **photosynthesis** by crops; and indirectly through climatic change.

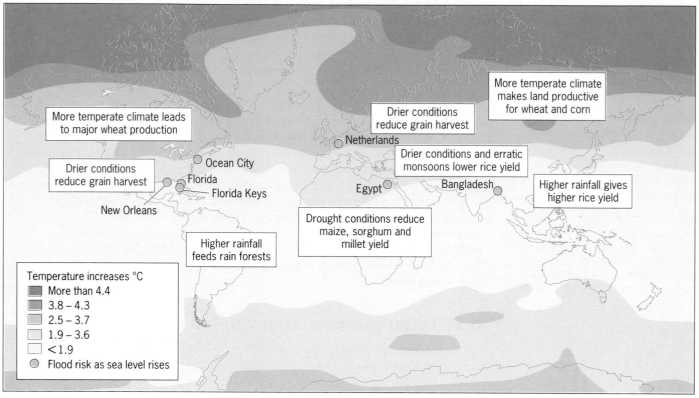

Figure 6.16 Average annual global temperature increases, until 2050

?

12 Study Table 6.4.

a Which greenhouse gases result from agricultural output?

b Draw diagrams to explain how agriculture contributes to the greenhouse effect.

Table 6.4 Relative contribution to the greenhouse effect of the various gases

Greenhouse gas	Source	Approximate relative greenhouse effect per molecule	Current rate of increase (% pa)
Carbon dioxide (CO$_2$)	Exchanged naturally between the atmosphere, oceans and the living world. The burning of fossil fuels and the loss of tropical rainforests adds to the normal amount.	1	0.5
Methane (CH$_4$)	Natural gas from rocks, swamps, rice fields and animals.	21	0.9
Nitrous oxide (N$_2$O)		290	0.25
Chlorofluorocarbons (CFCs)			
CFC11	Aerosols, fridges, blown	3500	4
CFC12	plastic packaging	7300	4

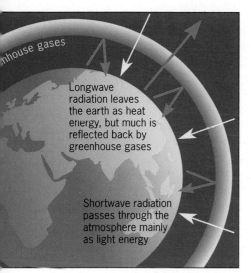

Longwave radiation leaves the earth as heat energy, but much is reflected back by greenhouse gases

Shortwave radiation passes through the atmosphere mainly as light energy

Figure 6.17 The greenhouse effect

Negative feedback

Figures 6.18 and 6.19 show the atmosphere as simple systems diagrams comprising inputs and outputs. Some scientists believe that the atmospheric system is self-regulating: that global warming may lead to greater evaporation, more cloud cover, and therefore greater reflection of incoming solar radiation and an equilibrium in temperatures. This chain of events which restores balance to a system is called **negative feedback**.

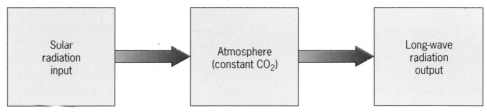

Figure 6.18 Model of climatic stability: input equals output so climate/temperature equilibrium maintained

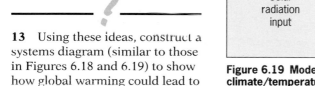

13 Using these ideas, construct a systems diagram (similar to those in Figures 6.18 and 6.19) to show how global warming could lead to negative feedback, which restores equilibrium to the global climate.

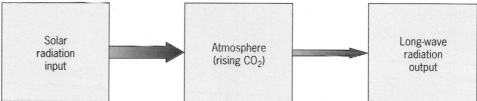

Figure 6.19 Model of climatic instability: input is greater than output so there is climate/temperature disequilibrium

Agriculture and increasing carbon dioxide levels

Rising levels of atmospheric carbon dioxide are, in fact, good for many crops. It has been suggested that the recent increase in carbon dioxide could increase rates of biological production by 15–20 per cent. Carbon dioxide assists photosynthesis and reduces water losses from crops. Essentially, it has a fertilising effect, causing certain crops to grow bigger and faster. Since these crops are mainly temperate, farmers in Europe and North America should benefit more than those in the tropics, provided there are no major changes in moisture availability. At current rates of carbon dioxide increase, it is estimated that wheat and barley yields might rise by 20 per cent by the year 2030. None the less, although the crops may grow better, there is still a risk of an increase in droughts.

Climatic change: winners and losers

Despite much research, we do not have an accurate view of what might happen to climate over the next 50 years. In fact, although the most important changes relate to precipitation, these are much more difficult to estimate than those for temperature. We must remember that the earth's atmosphere is fundamentally an unpredictable system, which makes even short-term forecasting difficult. None the less, various scenarios based on the greenhouse effect have been made and all of them point to a significant shift in climate. The impacts of change would be uneven (Fig. 6.16), and while some countries would benefit, some would lose out.

Global warming and the greenhouse effect

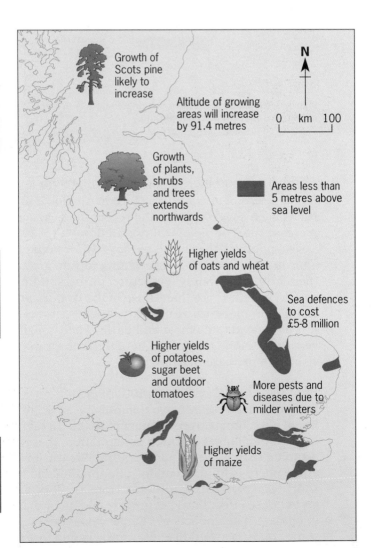

?

14 Draw a spider diagram to illustrate the climatic and economic pressures on a country such as Bangladesh (see also Sections 7.9 and 10.5).

Economically developing countries

In the economically developing world, the effects of global warming would be little short of disastrous. Already millions of people survive at subsistence levels in these countries. Thus any reduction in crop yields is likely to cause widespread food shortages. This can only add to the misery of people living in these areas. It is predicted that sub-tropical semi-arid areas, such as sub-Saharan Africa, would become drier, and yields of subsistence crops like sorghum, millet and maize would be drastically reduced. Rising temperatures would increase evaporation and lead to the abandonment of non-irrigated land and the expansion of deserts.

It is important to stress, though, that such predictions related to global warming exclude the problems caused by the separate issue of desertification (see Section 6.6). In Asia, a northerly shift of the monsoon belt would limit rice production in India, Bangladesh, Burma and Korea. Even more serious are the effects of rising sea levels. The projected rises of 20–40 centimetres by the end of the 21st century would threaten millions of hectares of unprotected farmland in deltas such as the Ganges-Brahmaputra in Bangladesh, and the Nile in Egypt.

Economically developed countries

Global warming could bring substantial advantages to farmers in the UK (Fig. 6.20); elsewhere climatic change will be less beneficial. In the US, a

Figure 6.20 The impact of global warming in the UK

Warmer climate could bring more pests and diseases. There could be large increases in insect pests such as aphids.

By mid-21st cnetury, the UK might have Mediterranean summers with mean annual temperatures of 14°C, compared to 11°C in the 1990s.

Winter warming would be greater than summer warming, though rainfall would increase, mainly from thunderstorms.

Crops such as maize and sunflowers, which in the 1990s are at their northern limit of ripening on the south coast of England, would be profitable further north.

Cultivation of native crops possible at higher altitudes.

Mean annual temperatures are forecast to rise by 0.5°C by 2000–2100; 1.5°C by 2020–2050; and 3°C by 2050–2100.

Growth of Scots pine likely to increase

Altitude of growing areas will increase by 91.4 metres

Growth of plants, shrubs and trees extends northwards

Areas less than 5 metres above sea level

Higher yields of oats and wheat

Sea defences to cost £5-8 million

Higher yields of potatoes, sugar beet and outdoor tomatoes

More pests and diseases due to milder winters

Higher yields of maize

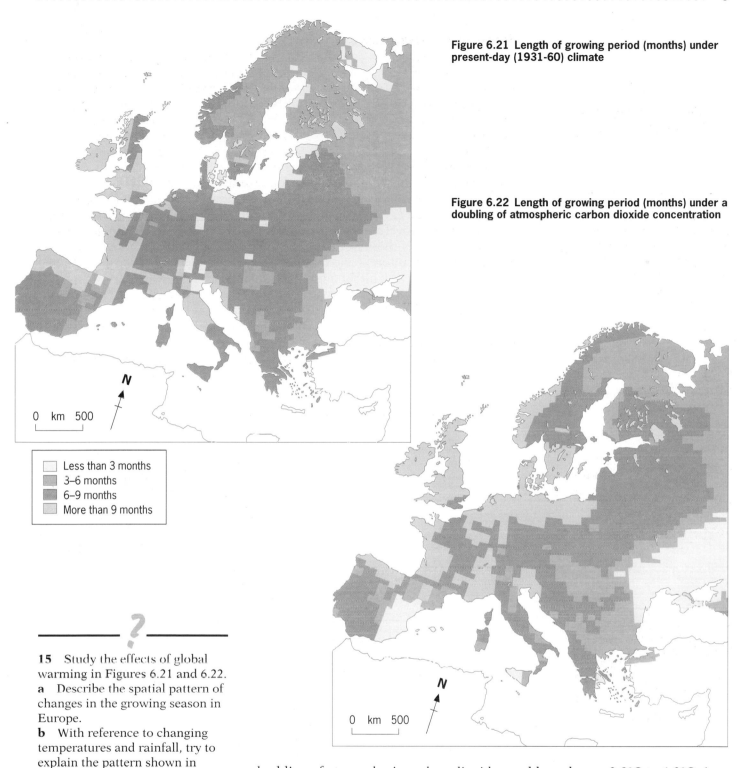

Figure 6.21 Length of growing period (months) under present-day (1931-60) climate

Figure 6.22 Length of growing period (months) under a doubling of atmospheric carbon dioxide concentration

0   km   500

Less than 3 months
3–6 months
6–9 months
More than 9 months

15 Study the effects of global warming in Figures 6.21 and 6.22.
a Describe the spatial pattern of changes in the growing season in Europe.
b With reference to changing temperatures and rainfall, try to explain the pattern shown in Figure 6.22.

16 Suggest ways in which farmers in: • the American Mid-West • semi-arid sub-Saharan Africa, might respond to climatic change. Why would their responses differ?

doubling of atmospheric carbon dioxide would produce a 3.8°C to 6.3°C rise in temperature. This might cause a 10 per cent reduction in soil moisture, and vast tracts of the American Mid-West, the greatest grain-producing region in the world, could become too arid for cultivation. The overall cost of reductions in crop yields is estimated at $33 billion a year, leading to rises in consumer prices and a massive fall in US food exports. This would have profound effects on many developing-world countries which rely on US cereal food aid to make ends meet.

?

17 Choose one of the agricultural impacts described in this chapter.
a Draw a 'vicious circle' diagram to represent the pressures making the impact worse.
b Annotate your diagram with actions needed to break the vicious circle.
c Comment on your diagram.

6.11 Responses to agricultural impacts on the environment

It is possible to find examples where both governments and individuals are making efforts to respond to environmental problems and manage the physical environment. One such effort is taking place in the UK with the designation of Environmentally Sensitive Areas (ESAs).

Environmentally Sensitive Areas

We have seen in this chapter that agriculture has had a profound effect on the environment. In the economically developed world, a return to a less intensive, organic agriculture may not be enough to repair the damage that has been done. Government intervention may be needed to encourage farmers, especially those in the most fragile environments, to adopt a **sustainable** agriculture – one that is managed in order to reduce damage to the ecosystem on which that agriculture depends.

A small step in this direction has been taken in the UK with the creation of Environmentally Sensitive Areas (Fig. 6.23). Farmers in ESAs receive payments for acting as custodians of the landscape (Fig. 6.24). In 1993, 17 000 farmers on 450 000 hectares were paid a total of £41 million to use

Figure 6.23 Environmentally Sensitive Areas in the UK

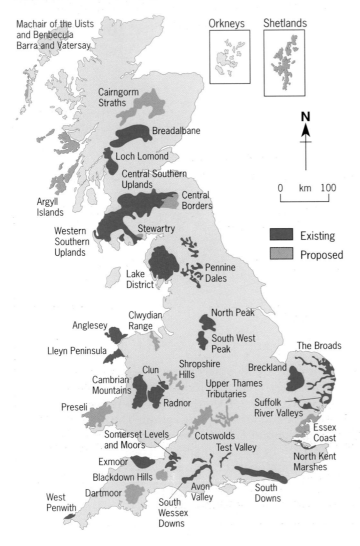

Table 6.5 Selected guidelines for farmers in the Pennine Dales ESA (*Source*: MAFF, 1992)

- Maintain grassland and do not plough, level or re-seed the land. This preserves the variety of plants in old, flower-rich meadows.

- Exclude livestock from meadows at least seven weeks before the first cut for hay or silage.

- Do not cut grass for hay or silage in any year before:
  1 July in Dentdale and Deepdale.
  8 July in Wharfedale, Langstrothdale or Waldendale.
  15 July in Swaledale, Arkengarthdale, Teesdale, Weardale and Rookhope.

- If you cut grass for silage, wilt and turn it before removal. This allows seed dispersal and prevents the disappearance of rarer species.

- Keep to your existing levels of inorganic fertiliser, provided it is below 20 units of nitrogen, 10 units of phosphate and 10 units of potash per hectare per year.

- Do not apply slurry or poultry manure. Slurry smothers herbage (pasture growth) and can seriously affect the composition of the sward (grass). Poultry manure can significantly increase soil fertility.

- Do not use more than 10 tonnes of farmyard manure per hectare per year, and apply in light dressings only. Too much fertiliser encourages the more aggressive plants which choke out the grasses and herbs. Many rarer plants cannot survive high nutrient levels in the soil.

- Do not use pesticides, fungicides and insecticides.

- Herbicides may be used to control bracken, nettles, spear thistle, creeping or field thistle, curled dock, broad-leaved dock or ragwort.

- Do not use lime, slag or any other substance to reduce soil acidity.

- Do not install any new drainage system or substantially modify any existing drainage system.

- You must maintain stockproof walls and hedges in stockproof conditions using traditional materials.

- Any weatherproof field barns which you own, or are responsible for, must be maintained in weatherproof condition using traditional materials.

- Do not damage or destroy any features of historic interest.

?

18 Study Table 6.5. Look back over this chapter and make notes about the environmental problems that an ESA designation may help to reduce. Evaluate the effectiveness of this policy.

19 Essay: Discuss the view that the British countryside is a cultural landscape worth preserving at any cost. In your scale of values where does conservation of the countryside come? Would you place it ahead of cheap food, a secure income for farmers, or the need for food self-sufficiency? Remember that conservation comes at a cost. What is your attitude towards this issue?

environmentally-friendly methods of farming (Table 6.5). Grants can reach £345 per hectare for work which varies according to the environmental issues in each ESA. In 1987 there were five ESAs, but this will have risen to 43 by the end of 1994.

Figure 6.24 ESA protection: mowing a traditional, species-rich (including orchids) hay meadow, Teesdale

Summary

- Agriculture has a profound impact on the environment.
- The environmental impact of agriculture may be deliberate or unintended, positive or negative.
- Environmental change often occurs in sudden spurts, after long periods of stability, and is caused by a wide range of economic, social, technological, political and cultural processes.
- The environmental impact of agriculture is often harmful and includes land degradation associated with soil erosion, salinisation, desertification and deforestation, as well as pollution.
- Over the next fifty years, global warming (caused by the enhanced greenhouse effect) could drastically alter world agriculture.
- Global warming might reduce world food production. Its effects will almost certainly be most severe in the economically developing world.
- At a regional scale the most disastrous effects of global warming would be felt in lowland coastal areas, where rising sea levels would flood huge areas of farmland.
- Significant climatic change (both temperature and rainfall) is likely to accompany global warming. Some farming regions could suffer, with many becoming too dry for cultivation. Western Europe, including the UK, is a probable exception. Here yields should rise, allowing the growth of a larger range of crops over a wider geographical area.
- Governments in the economically developed world have acknowledged the need for a balance between food production and environmental protection. In the UK the creation of Environmentally Sensitive Areas is a significant first step in this direction.

7 Factors of production and decision-making

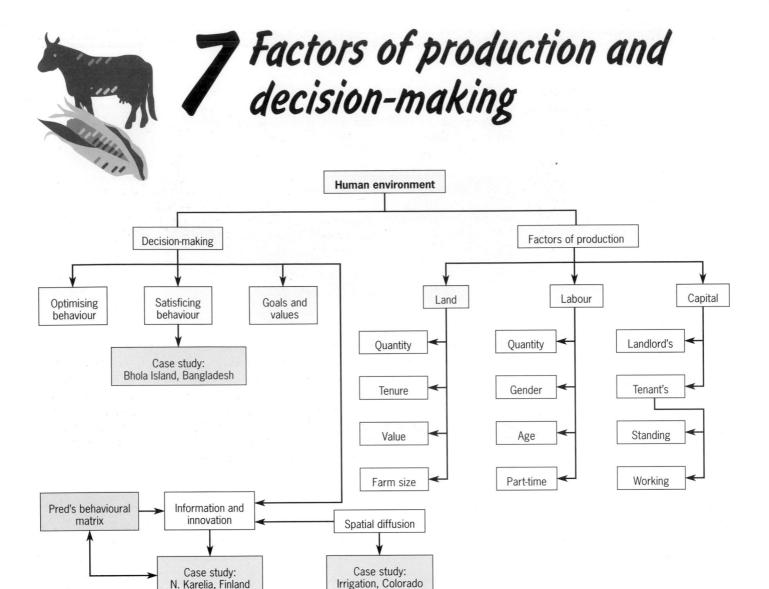

7.1 Introduction

Agriculture, like manufacturing and service industries, is an economic activity which relies on the basic factors of production. First we shall describe this basic economic framework, and then move on to consider how farmers operate as individuals in making decisions.

7.2 Factors of production

All farming enterprises depend on, in varying degrees, land, labour, capital and entrepreneurship. These are the factors of production. Land is needed for crops and livestock; labour, even on highly mechanised farms, is essential for cultivation; capital may be in the form of buildings, machines, tools, fertilisers and so on; and finally all farmers, from the Sahel of Africa to the Great Plains of North America, are entrepreneurs or decision-makers.

7.3 Land

There is more to agricultural land as a factor of production than simply its physical area or its fertility. We shall concentrate on just four of its characteristics: quantity, tenure, value and farm size.

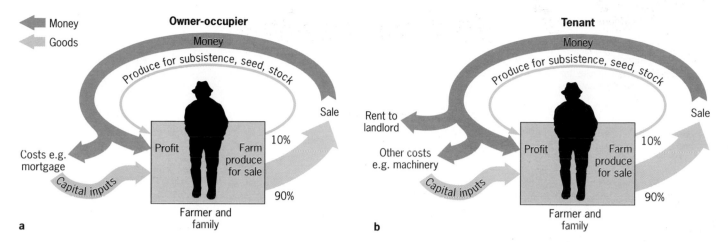

Money
Goods

Owner-occupier

Money
Produce for subsistence, seed, stock
Sale
Costs e.g. mortgage
Profit | Farm produce for sale
10%
90%
Capital inputs
Farmer and family

a

Tenant

Money
Produce for subsistence, seed, stock
Sale
Rent to landlord
Other costs e.g. machinery
Profit | Farm produce for sale
10%
90%
Capital inputs
Farmer and family

b

Figure 7.1 Characteristics of owner-occupier and tenant farm systems

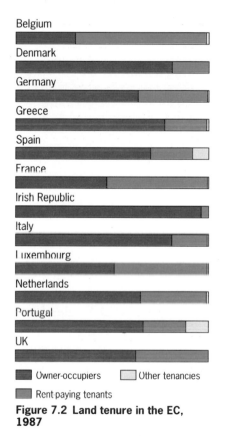

Belgium
Denmark
Germany
Greece
Spain
France
Irish Republic
Italy
Luxembourg
Netherlands
Portugal
UK

■ Owner-occupiers □ Other tenancies
■ Rent-paying tenants

Figure 7.2 Land tenure in the EC, 1987

?

1a For each of the forms of land tenure in Figures 7.1 and 7.3, suggest some advantages and disadvantages for the farmer.
b Draw a table to show these contrasts.
c Which system would you recommend? Give reasons for your choice.

Quantity

Since about 1960 in the developing world, there has been an expansion of cultivated land as the demand for food has risen with population growth. In many parts of Africa and South America this process accounts for up to 80 per cent of the increase in food **output**. However, in economically developed countries the picture is very different. There, as a result of urbanisation, the total agricultural area has fallen. For instance, in the UK, 471 000 hectares of agricultural land were lost (mainly to urbanisation) between 1976 and 1992. This represents an area equal to the size of Lancashire.

Tenure

Farmers may be either owner-occupiers (i.e. owning their farms) or tenants who pay rent to a landlord (Fig. 7.1). During the nineteenth century, the British countryside was dominated by large estates farmed by such tenants. In fact, in 1910 owner-occupiers controlled only 12 per cent of farmland in England and Wales. Gradually, though, this ownership pattern has been reversed, bringing the UK into line with other western European countries by the second half of the twentieth century (Fig. 7.2).

In many parts of the developing world, share cropping is a common form of land tenure. In a typical sharecropping arrangement in Bangladesh, a landlord would provide all of the capital **inputs** (seeds, fertilisers, ploughs etc.) for a tenant, and would then receive 50 per cent of harvested crops as rent (Fig. 7.3).

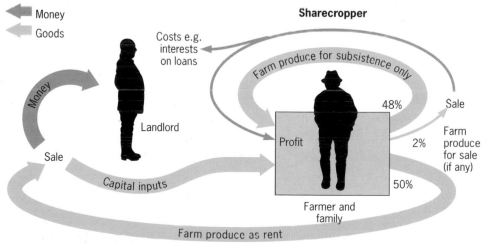

Money
Goods

Sharecropper

Costs e.g. interests on loans
Money
Landlord
Sale
Capital inputs
Farm produce for subsistence only
Profit
48%
Sale
Farm produce for sale (if any)
2%
50%
Farmer and family
Farm produce as rent

Figure 7.3 Characteristics of a sharecropping farm system

Table 7.1 Explanations of farm size (*After*: Bowler, 1992)

- Pressure of population. Countries with a low population density, such as Canada and Australia, have many large farms, whereas countries with a high population density, like Italy, Greece, Bangladesh and Indonesia, are dominated by small-holdings.

- Capitalism and communism encourage large farms. In capitalist systems, large farms, which gain maximum benefit from **economies of scale** and low unit costs, are more profitable than small farms. In communist systems, economies of scale are enforced as part of the state plan for agricultural output. The UK was the first European country to feel the impact of the capitalist process and in consequence has the largest farms in the EC. In the former USSR, state farms were very large and have been broken up as part of the on-going privatisation programme.

- Farms are usually larger on poorer land in order for the farmer to gain some level of profitability. This is illustrated by government land reforms in the Punjab (India), which created 7 ha farms on irrigated land capable of producing two crops a year, and 20 ha farms on non-irrigated arable land.

- Proximity to urban areas allows small farms to survive because their costs of farm inputs and marketing are relatively low.

- Government legislation may prevent the division of farms among surviving relatives on the death of the farmer. There has also been state-sponsored land reform (e.g. Cuba) which has either divided estates or has made large collective farms.

- In the developing world the sub-division of farms on the death of the landowner is still common. For example, in Islamic culture land is often divided equally among sons, with daughters getting a half share. Such inheritance practices result in tiny, fragmented farms, which may be too small for subsistence.

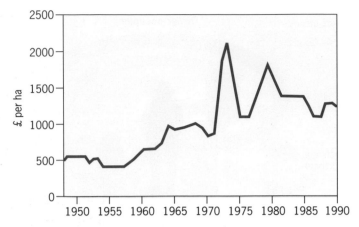

Figure 7.4 Agricultural land prices in England and Wales, 1948–90 (*Source*: MAFF/IR Land Price Series)

Table 7.2 Farm size and farming intensity in the EC, 1990–1

Enterprise	Agricultural area farmed (ha)	Total output (million ECU*)
Cereals	39.0	32.0
General cropping	19.8	32.3
Horticulture	3.5	91.6
Vineyards	7.8	39.2
Fruit and other crops	7.0	17.9
Dairy	28.4	66.2
Livestock	39.1	32.3
Pigs/poultry	10.9	176.7
Mixed	25.8	50.3

*ECU = European Currency Unit

2 Using the information in Table 7.2, test the hypothesis that farm size influences farming intensity.
a Work out the farming intensity of each enterprise in Table 7.2. You can do this by calculating the output per hectare of each enterprise.
b Plot the output per hectare (*y*) against the agricultural area farmed (*x*) as a scattergraph. Describe the relationship between the two variables.
c Now make a more precise analysis of the relationship between farm size and intensity using Spearman's rank correlation test (see Appendix A1).
d Comment briefly on your results and try to explain them.

3 Look at Table 7.3. Try to explain the apparent paradox of a declining percentage and an increasing number of rural population.

Value
Before World War 2, economic recession kept agricultural land prices low in Europe and North America. After the war, however, there was a steep rise in land prices. Whereas in 1943 one hectare of farm land in England cost on average £62, by 1991 its value had risen to £4198. Even allowing for inflation, this was a considerable increase (Fig. 7.4). Such a massive increase in value of the farmer's chief asset has been the key to much agricultural development in the UK in the post-war period.

Farm size
Farm size can be measured in several ways: as hectarage, amount of labour input, and as business size (capital investment). Table 7.1 gives some explanations for variations in farm size.

7.4 Labour

Traditionally, labour has been the most important factor of production in farming. As in manufacturing and service industries, farm labour varies in quantity, gender, age, full-time and part-time working.

Quantity
Agriculture continues to dominate the employment structure of most developing world countries (Fig. 7.5). For instance in Burkina Faso and Bhutan, over 85 per cent of the employed population works in agriculture.

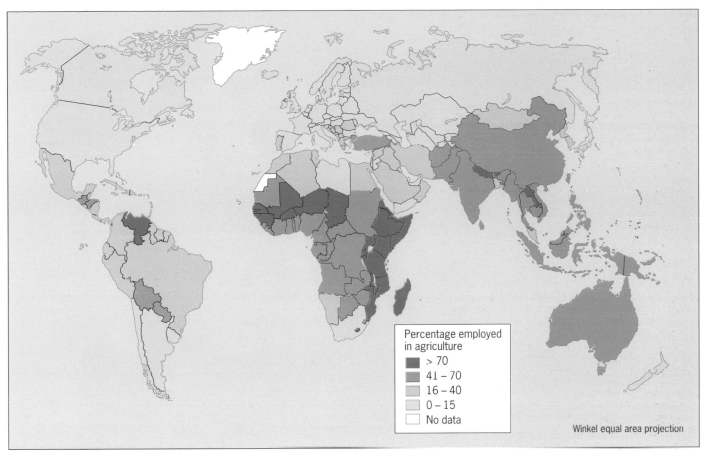

Figure 7.5 World employment in agriculture, 1990

Table 7.3 Bangladesh: changes in the rural population, 1970–90

	1970	1990
Percentage rural population	81	69
Total population	88 219 000	115 593 000

Figure 7.6 Farming family take a break from hay making, Yorkshire, UK

However, in the last 25 years there has been a substantial reduction in the percentage of people living in rural areas in the developing world. The reasons for this are not simple, though rural poverty is an important cause. You should also realise that a fall in the percentage of rural dwellers does not necessarily mean a fall in absolute numbers (Table 7.3).

'Push' and 'pull' factors

There is a strong relationship between the proportion of the workforce engaged in agriculture and levels of poverty in the developing world. Even so, our earlier studies of farming in Tamil Nadu (Chapters 2 and 4) and North Korea (Chapter 2) showed that labour requirements vary with farming type. For example, wet rice cultivation in South China and Java has a very high demand for labour compared to shifting agriculture in central Africa.

In the UK in 1993, only 2.5 per cent of the employed population worked in agriculture while under half of the labour force in the food industry lived in rural areas. The UK's agricultural workforce declined from around one million in 1981–3 to 890 000 in 1993. This change resulted from a combination of 'push' and 'pull' factors. Mechanisation continues to push farm labourers from the land, while higher wages and less strenuous work in manufacturing and service industries pull rural dwellers to urban areas.

There are full-time hired workers on only one-quarter of all holdings in England and Wales. The larger arable farms of eastern England employ most hired workers while smaller livestock enterprises in the North, West, Wales, Scotland and Northern Ireland get by with just family labour (Fig. 7.6). None the less, hired labour is still more important in the UK than in any other EC country (Fig. 7.7).

?

4a Using Figure 7.7, describe the geographical pattern of agricultural employment across the whole of the EC.
b Contrast this with the pattern shown for the UK.

5 Read through Figure 7.8.
a Why are the children of farmers often not interested in farming as a career?
b Where are the most stable, and the least stable farms? Suggest reasons for this pattern and say whether it is typical of the EC (Fig. 7.7).

6 While agriculture has fewer workers than ever before, its labour force has become more complex. Study the information in Figure 7.9 and write three more paragraphs for the newspaper article (Fig. 7.8) about the main changes in the UK's agricultural workforce between 1981–3 and 1992.

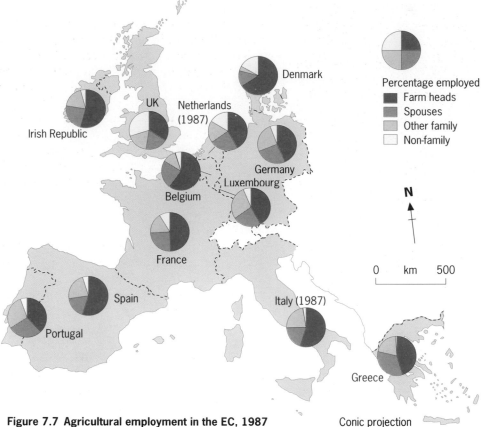

Figure 7.7 Agricultural employment in the EC, 1987   Conic projection

Farms 'left without future'

Many children of the present generation of farmers do not want the hard work and uncertainty of the job. According to a survey published yesterday, there is no one to take over almost half of farms.

The survey of the state of agriculture by the National Westminster Bank shows an industry riddled with doubts.

'There is clear evidence of a missing generation of farmers — sons of farmers either failing to see a viable future or simply finding the prospect of uncertainty unattractive,' the survey says.

The most stable farms are in the least prosperous areas — in the north of England, Wales and the west of Scotland. The most vulnerable are in the South-East, where 56 per cent of farms have no successor, and among the most affluent concerns, the very large arable farm.

Figure 7.8 Farming without a future? (*Source:* Maev Kennedy, *The Guardian*, 29 Feb. 1992)

Gender and age

Age

In economically developed countries, such as the UK, the workforce in agriculture is ageing: young people are reluctant to join what they see as a declining industry with few prospects (Fig. 7.8).

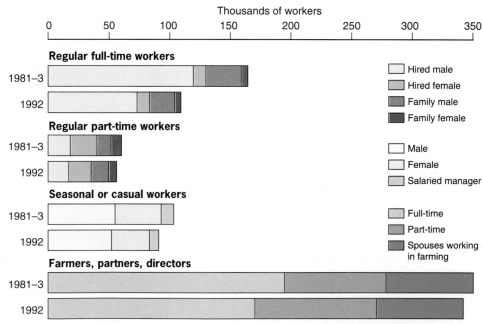

Figure 7.9 Agricultural labour force in England and Wales (*Source:* MAFF 1993a)

Women

Contrary to popular belief, on a global scale women have an important role in agriculture. In fact, in the some parts of the developing world they dominate certain farming sectors. In Africa, for example, it is estimated that women contribute 60 to 80 per cent of labour requirements in **subsistence** crop cultivation (Fig. 7.10). However, in the Islamic areas of rural Asia their situation is very different. Here tradition prevents women from working in the fields. Instead, adherence to the custom of *purdah* (keeping women veiled and in seclusion) restricts women to domestic tasks (cooking, sewing etc.), carrying water, and collecting fuelwood.

In economically developed countries such as the UK, women are less involved in manual labour but deal with much of the administrative side of farming (Tables 7.4 and 7.5).

Part-time farming

A large number of EC farmers are part-time, or even regard their farms as a hobby. In England and Wales over half of all holdings have a labour input of less than 275 Standard Work Days a year. In other words, there is not enough profitable work to occupy the farmer full-time.

Two-thirds of farm families in England and Wales have at least one source of off-farm income. This includes contracting work for other farmers, which is most popular, followed by holiday accommodation and horse livery. The opportunities for off-farm employment, however, vary regionally. They are highest in the South-East and lowest in the North, North-West and Wales. This dual employment is generally much stronger in the rest of western Europe than in the UK (Fig. 7.11).

Figure 7.10 Women tilling soil for planting, Zambia

Figure 7.11 Percentage of EC farmers with other gainful activities, 1987 (*Source:* European Commission, 1993)

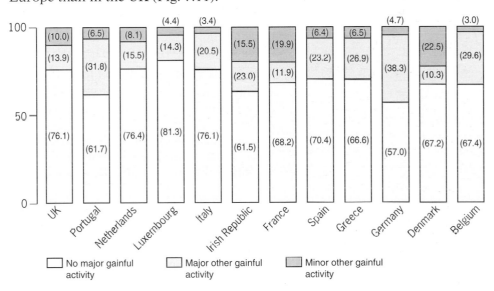

7 Study Figure 7.11.
a Describe the range in the proportion of EC farmers in:
- full-time farming,
- other major gainful activities.
b Describe the broad geographical pattern of part-time farming in the EC.
c Compare this with the pattern for agricultural employment from question **4a**. Comment on your findings.

8 Suggest how the proportion of part-time farmers in any country might:
a affect farming intensity and output per hectare.
b be affected by farm size.

Table 7.4 Farm wives' manual labour input in the UK, 1986–7 (*Source*: Gasson, 1992)

Farm type	Percentage of total input
Dairying	5.2
Upland livestock	7.5
Lowland livestock	9.0
Cropping	1.7
Pigs and poultry	5.2
Horticulture	7.3

Table 7.5 Farm wives who perform business tasks in the UK, 1986–7 (*Source*: Gasson, 1992)

Task	Percentage of farm wives
Dealing with callers, telephone	90
Running errands for business	72
Discussing farm business	67
Farm office work	67
Dealing with employees	27

7.5 Capital

Capital in agriculture means all of the materials and financial resources used for production. A distinction is made between a landlord's (or fixed) capital and a tenant's capital. The former includes land, buildings, roads and drains, while the latter includes standing capital (or farm assets), such as machinery and livestock (which are used repeatedly and are replaced only when they are worn-out or obsolete); and working capital such as seed, fertiliser, feed and wages for labour.

Commercial agriculture is often capital-intensive. In other words, it is heavy on machinery and light on labour, making investments in technology to boost output. This contrasts with subsistence agriculture where capital inputs (compared to labour inputs) are small. In **shifting agriculture** capital assets may consist of nothing more than hoes, axes and seeds. **Peasant farmers** usually have more capital (though share croppers may rely exclusively on capital provided by their landlord), and own (or have the use of) oxen, ploughs, carts and so on. However, peasant farming is essentially labour-intensive, rather than capital-intensive.

Farmers in the developed world are often said to be 'asset-rich but income-poor'. Compared to other businesses they are in an unusual position for, while they have large assets, their incomes are often disappointingly low. Most farmers used to finance their own capital requirements through re-investment of profits or from borrowing within the family. However, in the 1970s and 1980s, other sources of borrowing, especially banks, became more important. By 1991–2 the average debt of a UK farmer was £65 600. Interest payments on this debt amounted to 31 per cent of farm business income in 1985.

9 Draw annotated diagrams to illustrate the difference in capital inputs between commercial and subsistence agriculture.

7.6 Decision-making by farmers

In Section 7.1 we listed entrepreneurship as one of the factors of production. Farmers are entrepreneurs, who own and/or manage agricultural enterprises. This means that they have an important role as decision-makers. Decision-making is something that all farmers have in common, whether they are peasant farmers or commercial farmers; owner occupiers or tenants; large or small holders. In the remainder of this chapter we shall focus on decision-making behaviour and the factors which influence it.

7.7 Optimising behaviour

Until the 1950s, theories of agriculture assumed (for the sake of simplicity), that farmers were **optimisers** and behaved as '**economic wo/man**'. The optimiser concept assumes that as the decision-maker, the farmer has **perfect knowledge** and seeks to maximise profits or minimise costs in a wholly rational way. Indeed, this was one of the basic ideas of von Thünen's theory (Section 8.6). However, as we shall see, economic wo/man had little in common with decision-makers in the real world.

Although most farmers might wish to maximise their profits, many obstacles prevent them from doing so. First, information is not equally available to everyone. Second, in economically developed countries, there is often too much information (from television, radio, newspapers and magazines) for farmers to take in. And, third, farmers vary in their ability to use information effectively.

It is also true that farmers have goals in life which do not involve maximising profit. This is because many farmers may value security and

Table 7.6 Influences on farmers' perceptions

1 Personal factors: social background, education, age, income, health etc.

2 Psychological factors, in turn influenced by:
- cultural and community traditions.
- beliefs, values, goals.
- personality, e.g. attitude to risk, motivation for hard work.

3 The availability of information and the farmer's ability to use it, e.g. in the application of new technologies and other innovations.

4 Government policies, market prices, capital availability for investment.

5 Experience and past perceptions.

?

10a Compare Figures 7.12 and 7.13 and describe the differences between potential returns and actual returns to farmers in central Sweden.
b Suggest possible reasons for the differences you have described.

stability more highly than increasing revenue. Faced with the uncertainties of prices and the weather, they choose to minimise risks, preferring crops which guarantee a steady return to those which offer a higher but less predictable yield.

Farmers also show considerable inertia (they resist change) when choosing agricultural enterprises. The theories in Chapter 8 suggest that farmers can easily switch enterprises and grow those crops which give the greatest profits. In reality, farmers have capital tied up in specialist plant (e.g. a dairy parlour) and equipment (e.g. grain driers, combine harvesters) which makes it difficult for them to change. Thus, despite better economic prospects, a farmer might find it impossible to convert from dairying to arable farming. Meanwhile, environmental constraints, such as a cold climate or infertile soils, put further limits on the scope for change.

7.8 Satisficing behaviour

It is now acknowledged that a farmer's personal view or **perception** of reality reduces his/her choice of options (Table 7.6). Simon (1957) was one of the first to point out the weaknesses of the optimiser concept. He argued that decision-makers were **satisficers** rather than optimisers; and that they had 'bounded', rather than perfect knowledge. This means that a farmer may not maximise profits simply because s/he is prepared to accept, and is 'satisfied' with, less than optimal returns. In other words, when farmers strive to maximise profits, their lack of knowledge and/or inaccurate perception prevents them from achieving this goal. Our conclusion is clear: farmers' decision-making is rarely optimal.

Potential and actual yields, central Sweden
Julian Wolpert (1964) studied farmers' behaviour in central Sweden. He looked at the resources available to farmers, including soil fertility, climate, farm size, amount of capital, machinery and so on. From this he calculated the potential returns if farmers used their resources optimally and behaved like profit maximisers. He presented his results as a map (Fig. 7.12). He then drew a second map to show the actual returns achieved by farmers (Fig. 7.13). Wolpert concluded that farmers in central Sweden failed to achieve their potential.

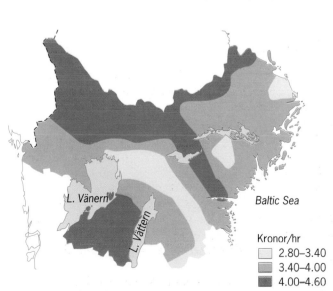

Figure 7.12 Optimal returns to farmers, central Sweden

L. Vänern

L. Vättern

Baltic Sea

Kronor/hr
- 2.80–3.40
- 3.40–4.00
- 4.00–4.60

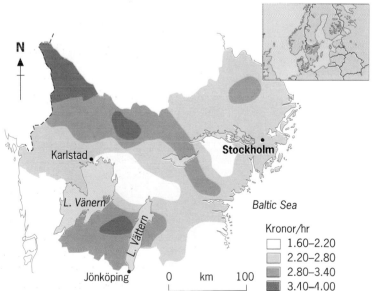

Figure 7.13 Actual returns to farmers, central Sweden

N

Karlstad

Stockholm

L. Vänern

L. Vättern

Baltic Sea

Jönköping

0   km   100

Kronor/hr
- 1.60–2.20
- 2.20–2.80
- 2.80–3.40
- 3.40–4.00

11 Study the data in Table 7.7.
a How accurate are farmers' perceptions of the drought hazard? Would you say that farmers were generally optimistic or pessimistic?
b What factors might influence the accuracy of a farmer's perception of drought?
c In a semi-arid environment, what might be the consequences of farmers' over-optimistic perception of drought? Draw a flow chart to illustrate your answer.

Drought perception in the Mid-West, USA

There is often a considerable difference between the actual frequency and magnitude of natural hazards like droughts and floods, and their perceived frequency by farmers. In the USA, drought is by far the biggest cause of crop losses. Saarinen (1966), in a classic study, analysed the drought hazard facing farmers in four Mid-West states. He questioned farmers from several counties where there was wide variation in the probability of drought. When he compared their perceptions with the actual drought frequency, he found that there were large differences (Table 7.7).

Table 7.7 The actual percentage of drought years compared to farmers' perceptions

State	County	Farmers' estimates	Reality
Nebraska	Adams	17.0	42.4
	Frontier	19.9	41.6
Kansas	Barber	16.0	46.9
	Finney	28.6	47.2
Colorado	Kiowa	34.9	47.2
Oklahoma	Cimarron	34.7	48.7

Perceptions of farmers in Bhola Island, Bangladesh

Farming features

Bhola is the largest island in the delta formed by the Padma, Brahmaputra and Meghna rivers in Bangladesh (Fig. 7.14). The physical resources for farming in this area are good. There are rich alluvial soils, abundant water supplies, and temperatures high enough (12°C – 38°C) to permit cultivation all year round. Rice is the principal summer crop and wheat is grown in winter. Farmers also grow various fruits, including coconut, mango and jack fruit (Fig. 7.15).

Figure 7.15 Coconut nursery, Bhola Island, Bangladesh

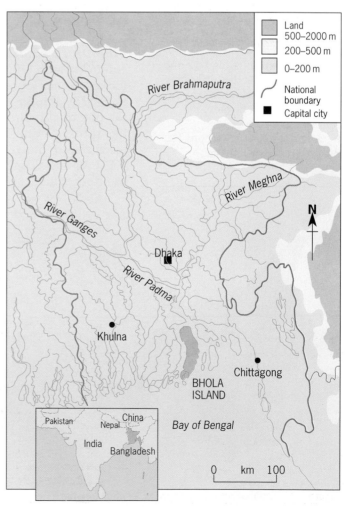

Figure 7.14 Bhola Island, Bangladesh

Yet despite the island's physical advantages, crop yields are low and 60 per cent of the population of 1.5 million lives below the poverty line. The reasons for this are complex, involving both physical circumstances as well as people's attitudes (Fig. 7.16). They include the fact that only a few people own land: most are tenants or sharecroppers. Population density is high (over 150 people per km²). The adult literacy rate is 20 per cent. About 80 per cent of the population are Muslim and 20 per cent are Hindu.

?

12 Use Table 7.6 and Figure 7.16. Give examples from Figure 7.16 for as many of the influences on the Bhola farmers' perceptions as you can.

13 Write a reply from a sharecropper to Mukul Rahman's report.

14 Comment on the role of perceptions in the decision-making of farmers on Bhola Island.

Figure 7.16 The Bhola people face many problems and these are also some of the main causes of their poverty: Extract from a report by the Bhola Project Director for ACTIONAID, a charity with development programmes in many countries (*Source:* Mukul Rahman, ACTIONAID, 1993)

Unemployment: Day-labourers do not get work every day. The ever-increasing disproportion between the workforce and the quantity of work not only encourages competition among the poor, but also helps the rich to buy their labour at very low cost.

Religious beliefs: Religion is still playing a key role in dominating the fate of the poor. They are made to believe since birth that everything on earth occurs the way the supernatural powers will it to. The women are discouraged from studying or working outside the home. The result is that the entire family is dependent on the wages of an individual male.

Dadan (leasing land or property): This forms a vital part of the chain of exploitation of the poor by the rich. The rich people give credit to cultivators at a very high interest rate. Sometimes, the entire earnings of the harvest go to pay the creditors.

Primitive method of cultivation: The peasants are still using a primitive method of cultivation. There is hardly any government initiative for excavating canals to facilitate water supply for irrigation. The peasants do not have any idea of upgrading the method of cultivation with the help of modern technology.

Economic hardship: These people are financially hard up, and they find it difficult to run even a small business with their petty amount of money. They spend their entire savings in one go when a member of the family falls sick. Very often, they do not have any options to properly utilise their earnings.

Dowry: The system of dowry reigns supreme during the settlement of a marriage. Amid the maladies that exist in the society, the dowry is one of the most bitter. The daughter leaves her parents with part of their property either sold or mortgaged. Once mortgaged, the property hardly ever comes back to the owners.

7.9 Goals and values

The goals and values of farmers are complex. One study of farmers in Cambridgeshire and the West Midlands suggested that farmers have similar goals, values and priorities (Table 7.8). Intrinsic values, including doing the work you like and independence, were ranked first, followed by instrumental values (achieving a satisfactory income), expressive values (meeting a challenge and self-respect from doing a worthwhile job), and finally social values. Again, you should note that these goals and values are far removed from the optimising behaviour of economic wo/man.

?

15 Use the ideas in Table 7.8 to design a questionnaire-based project on farmers' goals and values. You might consider:
a the extent to which farmers are optimisers (i.e. profit maximisers).
b the importance of behavioural factors like those in Table 7.8.
c how goals and values vary among farmers with age, farm size, land tenure etc.

Table 7.8 Dominant values in farming

Priority 1: Intrinsic
- Enjoyment of work tasks
- Preference for healthy, outdoor farming life
- Purposeful activity, value in hard work
- Independence, freedom from supervision, ability to organise own time
- Control in a variety of situations

Priority 2: Instrumental
- Making a maximum income
- Making a satisfactory income
- Safeguarding income for the future
- Expanding the business
- Providing congenial working conditions

Priority 3: Expressive
- Feeling pride of ownership
- Gaining self-respect from doing a worthwhile job
- Exercising special abilities and aptitudes
- Having opportunities to be creative and original
- Meeting a challenge, achieving an objective, personal growth

Priority 4: Social
- Gaining recognition, prestige as a farmer
- Belonging to the farming community
- Continuing the family tradition
- Working with other members of the family
- Maintaining good relationships with workers

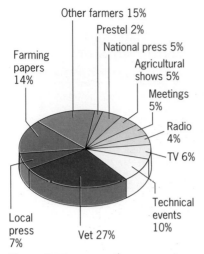

Figure 7.17 Information sources for UK farmers (*Source:* Fearne, 1990)

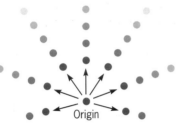

Figure 7.18 Point of origin: the effect of distance (neighbourhood effect) from origin of innovation

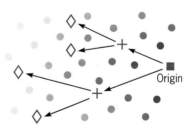

Figure 7.19 Hierarchy: individuals in larger places adopt information earlier than people in places lower down the hierarchy

Figure 7.20 Speed: the speed of adoption of innovation is affected by the efficiency of communications

Figures 7.18–7.20 Main characteristics of spatial diffusion of information

Figure 7.21 Curve of adoption of innovations by farmers

7.10 Information and innovation

Farmers receive information about new crop varieties, new pesticides, new types of machinery, and changes in government policies from many different sources (Fig. 7.17). Talking to neighbours or relatives (who are often trusted and respected) is a valuable information source. Similarly, newspapers, magazines and television are also popular, and in some countries professional advisers (often employed by the state) introduce new ideas to farmers. Finally, the representatives of commercial companies are also important.

This flow of information has a number of features (Figs 7.18–7.20):

1 It is a spatial process in which distance from the point of origin is an important control.

2 It is hierarchical: in other words, information trickles down from the Ministry of Agriculture (or other such organisation) through various channels until it reaches individual farmers.

3 It varies in speed according to the media. Thus television and radio spread information more rapidly than newspapers or magazines.

Of course farmers also vary in their receptiveness to information and new ideas (Fig. 7.21). Go-ahead farmers or innovators are often more willing to try out new ideas. We refer to these farmers as 'early adopters'. Compared to late adopters (or laggards), this group often has larger farms and more specialised enterprises; they are better educated; they are often younger and more willing to take risks; and they are better connected, often having a higher social status.

Consequently, the decision-making process can be broken down into a succession of stages (Fig. 7.22): awareness – where farmers first learn about a new idea; assessment – where farmers consider the idea new to them; trial – where farmers try out the idea; and finally adoption of the idea.

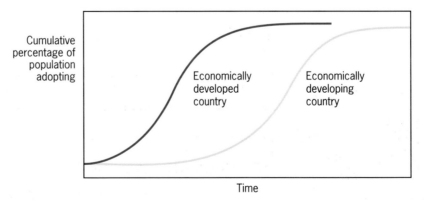

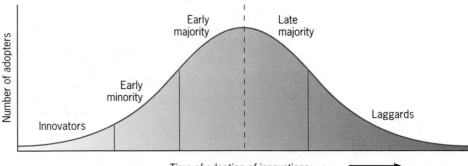

?

16 The adoption of hybrid corn in Iowa between 1926 and 1942 went through the stages of awareness, assessment, trial and adoption. Study Figure 7.22 and complete the following exercises.

a State three ways in which the behaviour of the innovators or early adopters differed from the other groups of farmers.

b Draw an adoption curve (similar to Fig. 7.21) to show the likely pattern of adoption of hybrid corn in Iowa. How does your curve differ from Figure 7.21?

c Suggest reasons why the laggards were so late in adopting hybrid corn.

17 Study Figure 7.23.

a Describe the pattern of adoption of wheat varieties between 1977 and 1986.

b Suggest two reasons for the wave-like pattern of adoption in Figure 7.23.

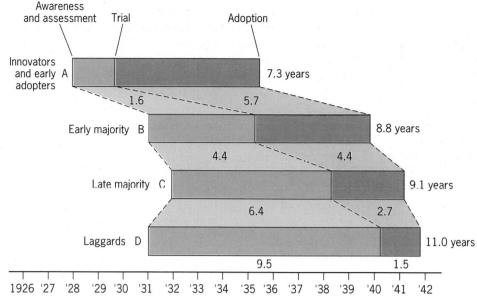

Figure 7.22 Adoption of hybrid corn in Iowa, USA: stages in the decision-making process

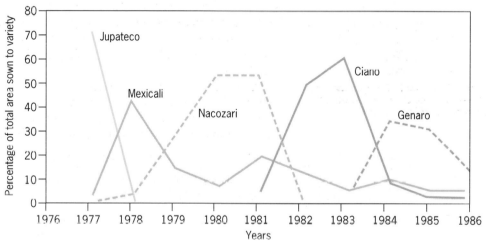

Figure 7.23 Adoption of wheat varieties in Yagui Valley, Mexico

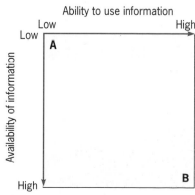

Figure 7.24 Pred's behavioural matrix

Pred's behavioural matrix

Alan Pred (1967) devised what he called a behavioural matrix (Fig. 7.24) to explain decision-making. His matrix used two factors: the amount of information available to decision-makers; and their ability to use the information. Pred argued that better-informed and more-able farmers would *probably* make the most successful decisions. Thus farmer B in Figure 7.24 is more likely to make a successful decision than farmer A. However, Pred understood that decision-making was an uncertain process. Thus, although the odds seem to be stacked against farmer A, s/he could none the less by chance make a good decision. Equally farmer B, with all his/her advantages, could make a disastrous decision. In this world of probability, nothing is certain.

The behaviour of farmer-sawmillers in northern Karelia, Finland

Pred's matrix has been used to study the behaviour of farmers in northern Karelia, Finland (Fig. 7.25). This region is close to the northern margin of cultivation for temperate crops, where severe environmental conditions make it difficult for farming (Fig. 7.26). Winters are long and cold, so the growing season is short. Poor drainage is another problem, especially in spring when the heavy snow cover melts. The soils, based on hard crystalline rocks, have limited farming potential (Fig. 7.27).

Northern Karelia is a region of small farms and this adds to the problems of farming. Most of the farmers here are part-time: they have dual occupations in forestry and sawn timber, as well as agriculture. Selby (1987) studied the behaviour of farmer-sawmillers in northern Karelia. Using a complex statistical technique he measured the farmers' ability and their access to information and plotted the two variables in a behavioural matrix (Fig. 7.25). Farmers in group 1 operated their sawmills for less than one month a year, while those in group 2 had greater commitment.

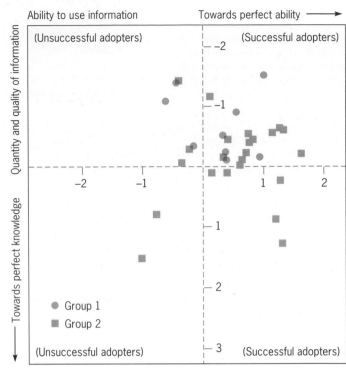

Figure 7.25 Behavioural matrix for farmer-sawmillers in northern Karelia, Finland (Selby, 1987)

Figure 7.26 Northern Karelia, Finland

Figure 7.27 Typical farming landscape in northern Karelia, Finland

?

18 With reference to Figure 7.25:
a State the main differences between the farmers in group 1 and those in group 2.
b Comment on the overall ability of farmers, and the availability of information. How does this compare with the concepts of optimising and satisficing behaviour?
c Which appears to be more important in successful decision-making: ability or information? Give reasons for your answer.

7.11 Spatial diffusion of agricultural innovations

If we map the spatial pattern of adoption of agricultural innovations over time, two influences become clear. First, there is a **neighbourhood effect**, with distance from some point of origin being a key control. And, second, physical factors often limit the innovation to areas with suitable climatic and soil conditions. The following case study illustrates the influence of these factors.

Irrigation in Colorado, USA

Conditions

Physical geography puts strict limits on dryland farming and cattle grazing in the northern High Plains of Colorado (Figs 7.28–7.29). With a growing season of only 145 days and frequent frosts, crops such as cotton and tree fruit are out of the question. Wheat cultivation is possible, but rainfall is low (430 mm a year on average) and unreliable. However, farmers can overcome the rainfall problem by using **irrigation**. This is costly though, for in spite of large groundwater resources across the area, they are at considerable depth. Any irrigation therefore requires expensive pumps.

Wells

In the mid–1940s a few innovative farmers installed the first wells. Between 1948 and 1962 the number of wells increased rapidly, from 41 to 410. Bowden (1965) mapped the spread of well irrigation during this period (Figs 7.30–7.32) and tried to predict the

future pattern if the adoption of irrigation continued unchecked. He believed that irrigation experiences swapped between farmers were crucial to understanding the spread of information about installing wells.

Mapping innovation

The first step in building the model was to estimate how easily information could be transferred within the state. Bowden used data on attendances at local barbecues and telephone calls for this purpose. This enabled him to predict the likelihood of contact between farmers and therefore the probability of a successful irrigator telling a friend or neighbour about the innovation. A computer simulation predicted the 1990 pattern of adoption (the only restriction was that no township could have more than 16 wells, as groundwater resources are finite). Bowden predicted a four-fold increase in the number of wells by 1990 (Fig. 7.33).

Figure 7.28 Dry, mountainous conditions in the northern High Plains, Colorado, USA

Figure 7.29 Northern High Plains area, Colorado, USA

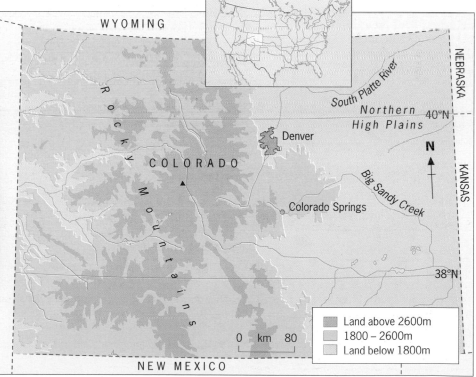

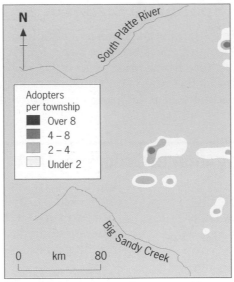

Figure 7.30 Irrigation wells per township: actual pattern, 1948

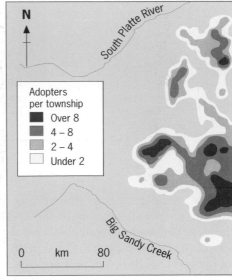

Figure 7.31 Simulated pattern of 410 wells

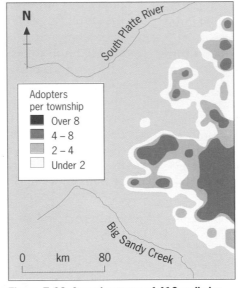

Figure 7.32 Actual pattern of 410 wells by 1962

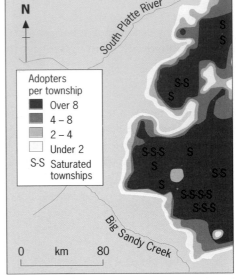

Figure 7.33 Simulated prediction of wells for 1990

?

19a How accurately does the simulated pattern of irrigation wells in Colorado fit the actual pattern (Figs 7.30–7.33)?
b What factors might explain differences between the simulated and actual patterns?

?

20 Essay: Outline the main factors influencing decision-making in agriculture, and assess their relative importance.

Summary

- There are four factors of production in agriculture: land, labour, capital and entrepreneurship.

- In all farming systems the factors of production set the basic economic environment for agricultural activity.

- Farmers are important decision-makers. Their decision-making behaviour suggests that they are satisficers rather than optimisers.

- Decision-making is affected by many behavioural factors, including quantity and quality of information, the ability of decision-makers, their experience, perceptions, goals and values.

- Distance has an important influence on the spread of innovations in farming.

8 Transport, markets and agribusiness

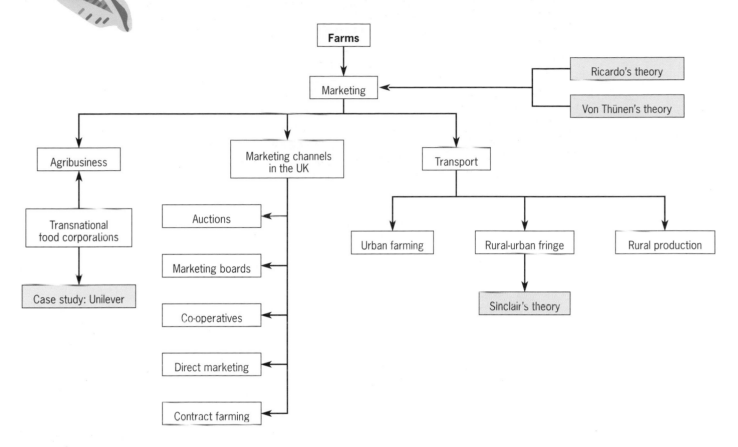

8.1 Introduction

In this chapter we extend our discussion of human factors to cover transport, markets and **agribusiness**. These factors are most relevant to **commercial agriculture**, which is geared to market demand and relies on efficient transport systems to link farmers with markets. Increasingly, transnational food corporations dominate commercial agriculture. They are associated with a large-scale, capital-intensive farming called agribusiness. As a farming **system**, agribusiness has more in common with manufacturing industry, than with traditional agriculture in economically developing countries (Dixon, 1990).

8.2 Transport

An efficient transport system is essential for modern commercial agriculture. Transport is needed to bring bulky items such as seed, fertiliser and agro-chemicals cheaply to the farm, and to distribute food products to markets. Some food products are both perishable and bulky and require transport which is cheap, rapid and/or refrigerated. A classic example is the British dairy industry which delivers liquid milk from farms to people's doorsteps in a matter of hours (Fig. 8.1).

Improved transport has brought about three major changes in food systems since the mid-nineteenth century. First, perishable products no

Figure 8.1 Milk tanker collecting milk from a farm dairy parlour for delivery to a home distributor

97

Figure 8.2 The distribution of registered cowsheds in London, 1881 (*Source:* Atkins, 1977)

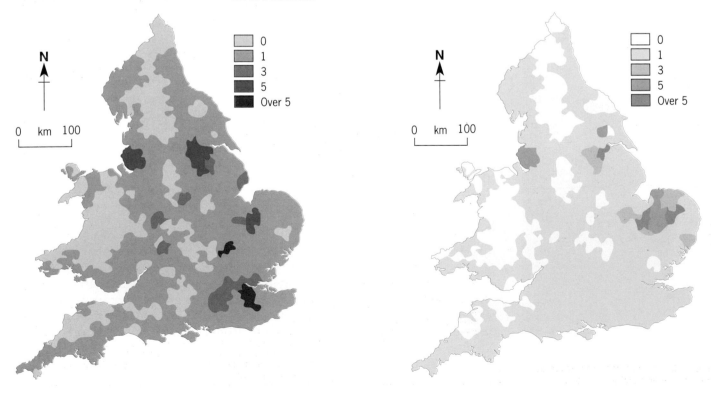

N

0 km 3

· Registered cow sheds

▨ Built-up areas

■ City centre

River Thames

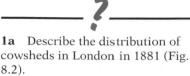

1a Describe the distribution of cowsheds in London in 1881 (Fig. 8.2).

b Suggest two possible factors which might explain the distribution.

2 Study Figures 8.3 and 8.4. Why might transport be an important factor influencing the distribution of the crops shown here? What other factors might explain the distributions?

longer have to locate close to the market. As late as 1881, much of London's milk supply was still produced in dark, crowded cowsheds within the city (Fig. 8.2) until improvements in rail transport and refrigeration allowed production to move away from the capital. Second, transport improvements encouraged regional specialisation based on the comparative advantage of climate, soil or organisation, rather than distance from market. Figures 8.3 and 8.4 show some of the regional specialisms found in the UK before joining the EC in 1973. (Since 1973 some of these specialisms have changed in the face of competition from European producers.) And third, as we have seen, improved transport allowed for the first time large-scale international trade in foodstuffs.

Figure 8.3 Beetroot as a percentage of the total grown in the UK (*below*)

Figure 8.4 Carrots as a percentage of the total grown in the UK (*below right*)

N

0 km 100

| 0 |
| 1 |
| 3 |
| 5 |
| Over 5 |

N

0 km 100

| 0 |
| 1 |
| 3 |
| 5 |
| Over 5 |

8.3 Agriculture and urban areas

Urban areas often provide markets for locally-grown farm products and may influence agricultural enterprises and agricultural land use in nearby rural hinterlands (see Section 8.6). In addition, urban areas may come into conflict with agriculture in the **rural-urban fringe**. The negative impact of urban areas on agriculture in this zone is the subject of R. Sinclair's model.

Sinclair's theory of farming in the rural-urban fringe

R. Sinclair (1967) said that farmers who located close to an expanding city in economically developed countries, would reduce their investment in agriculture. In this rural-urban fringe zone, agriculture would have to compete for land with commerce and industry. Because these other land uses could make higher profits than agriculture, they would eventually gain control of the land. Thus, sooner or later farmers in the rural-urban fringe expect to sell their land to urban uses. In the mean time, there is little incentive for them to invest in the land or farm **intensively**. As a result, farmland in the fringe belt may be left idle or cultivated **extensively**, with only limited **inputs** per hectare.

Sinclair argued that this process would cause land values and farming intensity to increase with distance from urban areas (Fig. 8.5). In theory, this should produce concentric rings (or zones) of agricultural land use and farming intensity around a city (Fig. 8.6).

Figure 8.5 Intensity of agriculture on the urban fringe according to Sinclair (*Source:* Ilbery, 1992)

Zone A: built-up area
In economically developed countries this is usually without any agricultural activity: in cities in economoically less developed countries, small livestock such as chickens are commonly kept.

Zone D: urban shadow
Urban areas will cast a non-visible shadow here in the form of land prices and land owner-ship patterns. But the outward appearance of the countryside is untouched.

Zone E: rural hinterland
Apart from some long-distance commuters and weekend urban visitors, this zone is not affected by the city.

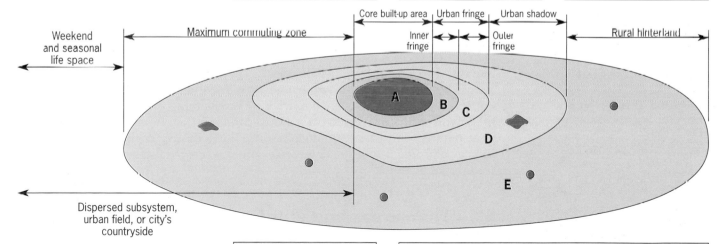

Zone B: inner fringe
An area of active conversion of farm land to urban uses, e.g. housing estates, parks, golf courses, hospitals, schools, crematoria, etc.

Zone C: outer fringe
Countryside already penetrated by some non-farming activities. Small farming hamlets and villages may have been swollen by commuter housing estates and some old farm buildings may have been converted to industrial and service uses. Owner farmers may be eager to sell their land for housing, in which case, there will be minimal investment in the land. Other farmers may intensify their operation and 'mine the soil' of its fertility before the onset of urbanisation.

Figure 8.6 Impact of urban areas on agriculture in nearby rural areas

Sinclair's theory

3 Look at Figure 8.7.
a What is the pattern of agricultural land use around the city?
b Compare this with Sinclair's model for agricultural location (Figs 8.5 and 8.6).
c Comment on your findings.

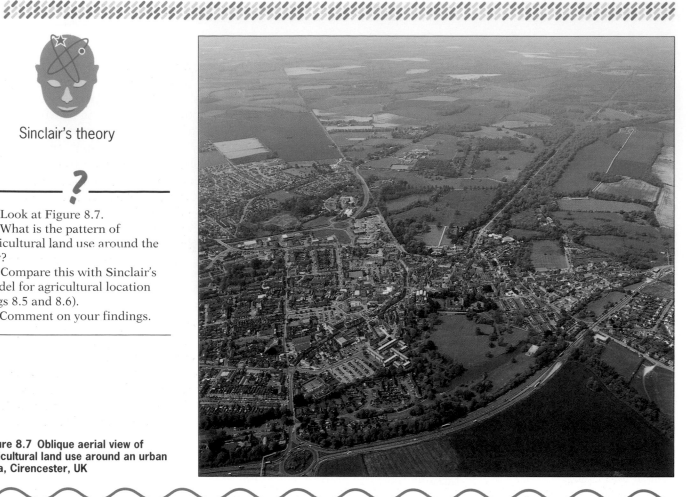

Figure 8.7 Oblique aerial view of agricultural land use around an urban area, Cirencester, UK

4 Investigate the influence of urban areas on agricultural land use by testing the following hypothesis: there is no significant difference in agricultural land use between the rural-urban fringe and the deep countryside.
a Collect sample land-use data either through fieldwork or from a 1:25 000 Land Utilisation map.
b Use chi squared (Appendix A2) to test the statistical significance of your data.

8.4 Farming in the rural-urban fringe

In reality, farming in the rural-urban fringe in economically developed countries is quite different from farming in the deep countryside. 'Horsiculture', for example, is common, reflecting the demand by urban dwellers for horse riding within a short distance of their homes. Agriculture in the fringe zone is also affected by planning controls. For example, in the UK, green belts often preserve agricultural land which would otherwise be swallowed up by urban growth. On the other hand, being near to large urban areas often creates problems for farmers such as vandalism, theft of crops, and the worrying of livestock by dogs.

Just beyond the rural-urban fringe, farmers often respond to the opportunities of a large urban market by intensifying production. Sometimes this involves **off-land enterprises** in agro-industrial complexes which house pigs, poultry, dairy cattle, and glasshouse horticulture. With their highly intensive production these enterprises often resemble factories rather than traditional farms.

8.5 Urban areas and employment in agriculture

In urban areas better job prospects and higher wages in industry and services create a competitive labour market. This affects employment in agriculture. As labour therefore becomes scarce and more expensive, farmers may consider investing in labour-saving machinery. It seems that mechanisation in British farming followed just such a pattern. Although the proportion of the workforce in British agriculture had dipped below 50 per cent before the end of the eighteenth century, it was not until 1851 that absolute numbers started to fall. The mechanical reaper for example, was available in the 1830s, but it was not widely adopted until the 1880s, when labour shortages became serious.

8.6 Markets

At the outset, you should appreciate that the term 'market' has several meanings: markets include not only individual consumers, shops, and centres of population, but also food processing factories, wholesalers and giant transnational food corporations, such as Nestlé and Unilever.

The location of markets influences spatial patterns of land use and farming intensity. This is especially the case in the economically developed world where commercial agriculture is dominant, and is primarily a response to market forces.

Agricultural land use and markets

Many factors – both human and physical – determine spatial patterns of agricultural land use. This makes these patterns extremely complex. To help us understand this complexity, geographers and economists have devised a number of simple models and theories. Two of the earliest theories are by Ricardo and von Thünen. Their general approach has been to isolate the effect of a single factor such as the market. By keeping all of the other factors constant and allowing this one factor to vary, they are able to demonstrate its impact on land use.

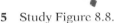

5 Study Figure 8.8.
a Why are rent levels highest on the most fertile soils?
b Which crop will be grown on:
• highly fertile,
• moderately fertile soils?
c Why is sheep farming (rather than crop cultivation) confined to soils of low fertility?

Ricardo's theory

David Ricardo, writing in 1817, said that the types of agriculture found in any place would depend upon either:
• the rent asked by the landlord (in the case of a tenant farmer), or
• the purchase price of the land (for an owner-occupier). The value of land affects land use because some crops/livestock are more productive than others. Thus, on high-value land, a farmer must grow high-value crops to make a decent profit.

Ricardo went on to argue that two factors are primarily responsible for agricultural land values. First there is local population density: where population density is high there is keen competition for land, which forces up prices. And second, there is land quality. Good quality farmland, with fertile soils, favourable climate and relief, would have a higher value than poor-quality farmland (Fig. 8.8).

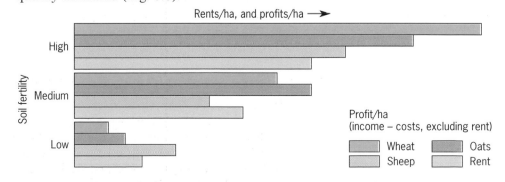

Figure 8.8 Ricardo's theory of agricultural land use

Von Thünen's theory

Another early theorist was Johann Heinrich von Thünen. He was born into a Frisian land-owning family and farmed 400 hectares near Rostock in Mecklenburg (Germany) on the Baltic coast. Von Thünen was a practical farmer rather than an economist and based his ideas on his own farming experience. He published these in his book, *The isolated state*, in 1826.

In the early nineteenth century Mecklenburg was a backward and isolated part of Germany; devastated by wars, it had only just converted from a medieval three-field system (see Section 6.2) to modern multi-crop rotation. Transport was slow, difficult and expensive.

Von Thünen wanted to examine the effect of the market on spatial patterns of agricultural production. His theory shows us how land-use patterns around a market *should* be organised to maximise profits for farmers. However, to make his theory workable, and to isolate the influence of the market, von Thünen simplified reality by making the following assumptions:

1   Decision-making is in the hands of farmers who are rational **economic wo/men**. Their goal is to maximise profits; they have perfect knowledge (i.e. they are fully informed about the needs of the market and the relative prices of goods); and they have perfect ability to use their knowledge.

2   The area farmed is an isolated state which is completely self-sufficient in agricultural produce.

3   The market is a single town, at the centre of a plain of uniform fertility.

4   The transport of agricultural produce to market is organised by farmers.

5   The market price for a particular product is the same for all farmers and does not vary over time.

6   The only transport is horse and cart and transport costs are proportional to distance and weight. Transport is equally easy in all directions.

7   All farms are the same size and there are no **economies of scale** available (see Section 2.4).

Using these assumptions von Thünen developed two models: a cropping model, and an intensity model.

The 'simple' model

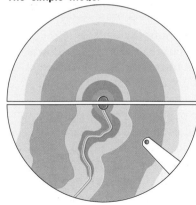

The model with two modifications

● Urban market ～ Navigable river

■ Free cash cropping (horticulture and dairying)

■ Forestry (wood)

□ Crop alternation system (six-year intensive arable rotation)

■ Improved system (seven-year rotation of arable with fallow and pasture)

□ Three-field system (arable rotation)

□ Stock farming (cattle and sheep grazing)

Figure 8.9 Von Thünen's cropping model: the isolated state (*After:* Von Thünen)

The cropping model

Von Thünen said that, with increasing distance from a central market, it would pay farmers to grow different crops. The outcome would be a series of concentric rings or zones of land use around the market (Fig. 8.9).

The **locational rent** of each crop determines this pattern of land use. Locational rent is the difference between the market value per hectare for a crop and the costs per hectare of cultivation and transport to market. We can summarise this with a formula:

$$R = E(p - a) - Efk$$

where R = locational rent per hectare
k = distance from the market (km)
E = yield of crop (tonnes/hectare)
p = market price (£ per tonne)
a = cultivation cost (£ per hectare)
f = transport cost (£/tonne/km)

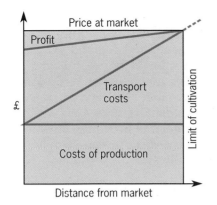

Figure 8.10 Von Thünen's theory: costs and distance from market

Because transport costs increase with distance, locational rent *decreases* away from the market. Eventually a point is reached where the locational rent of a crop is zero and cultivation becomes unprofitable. This point marks the spatial limit of cultivation of that crop (Fig. 8.10).

Each crop has its own unique locational rent curve. This is because crops vary in their market values, yields, cultivation and transport costs. If we plot several locational rent curves for different crops on a graph (Fig. 8.11) and rotate them through 360 degrees, the result is a series of zones. It follows that the land use in each zone is the one with the highest locational rent at that distance.

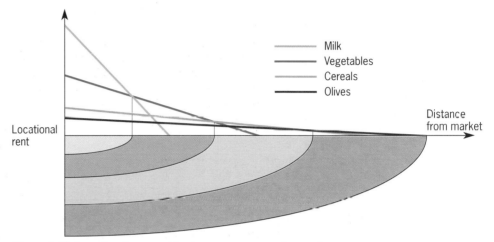

Figure 8.11 Locational rent and land-use zones around a market

Table 8.1 Determining agricultural land use patterns: market prices and costs

	Potatoes	Wheat
Market price (£/tonne)	100	125
Yields (tonnes/ha)	40	6
Transport costs (£/tonne/km)	1	1
Cultivation costs (£/ha)	250	250

Table 8.2 Calculation of locational rent

Distance to market (km)	Cultivation costs		Transport costs		Total costs		Market price		Locational costs	
	P	W	P	W	P	W	P	W	P	W
1	250	250	40	6	290	256	4000	750	3710	494
10	250	250	400	60	650	310	4000	750	3350	440
50										
100										

6 Assume that two crops are grown around a central market: potatoes and wheat. Conditions are those specified in von Thünen's theory. Crop yields, market prices, transport costs and cultivation costs are given in Table 8.1.

a Using the information in Table 8.1, complete the calculations of locational rent for potatoes and wheat in Table 8.2.

b Plot the locational rent (*y*) for potatoes and wheat against distance from the market (*x*) on a graph.

c At what distance from the market will potatoes give way to wheat?

The intensity model

This part of von Thünen's model stresses the link between crop yields, inputs such as labour and fertiliser, and location. Because rents near to the market are high, von Thünen argued that farmers in this area would intensify their inputs (note that this is the exact opposite to Sinclair's argument in Section 8.3). This would give higher yields which would cover the extra cost of land. According to the model, both dairying and market gardening should be found close to the market: dairying because milk is highly perishable; and market gardening because crops such as strawberries, lettuce etc. are not only perishable, but easily damaged in transit.

Von Thünen anticipated a 30 kilometre radius for the dairying and market gardening zone, its outer limit being set by the distance over which manure, street sweepings, and other town waste could profitably be transported to enrich the land.

Putting von Thünen to the test

If von Thünen's ideas are correct, we might reasonably expect to find evidence of rings of agricultural land use around major cities in the nineteenth century and in the economically developing world today.

Von Thünen's theory

Farming around London in the nineteenth century

London in the mid-nineteenth century seemed to fulfil the main conditions of von Thünen's theory. It was a huge market which generated demand for a wide range of agricultural commodities, while transport by horse and cart was slow and expensive. However, the evidence for zonal patterns of market gardening is disappointing. Figure 8.12 shows that market gardens clustered alongside the River Thames rather than around the urban fringe. Von Thünen did allow for the distorting effect that a waterway providing cheap transport might have (in this instance it made it possible to transport manure cheaply), but the main attraction of the Thames Valley was its fertile and freely draining loams and gravel soils.

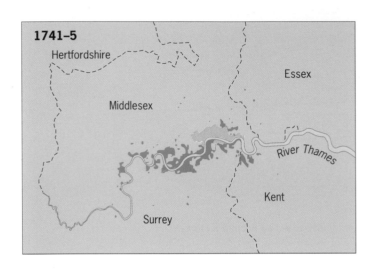

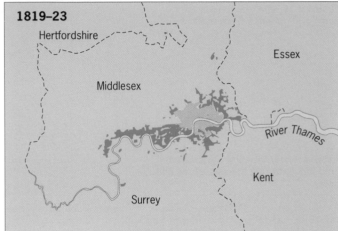

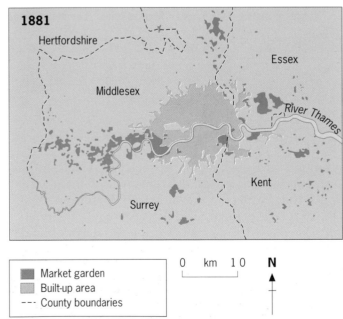

Market garden
Built-up area
--- County boundaries

0 km 10 N

Figure 8.12 Market gardens near London (*Source:* Atkins, 1987)

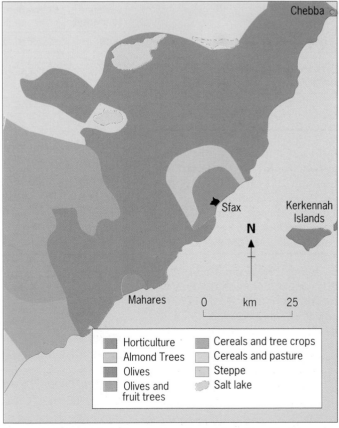

Horticulture
Almond Trees
Olives
Olives and fruit trees
Cereals and tree crops
Cereals and pasture
Steppe
Salt lake

Figure 8.13 Agricultural land use patterns near Sfax, Tunisia

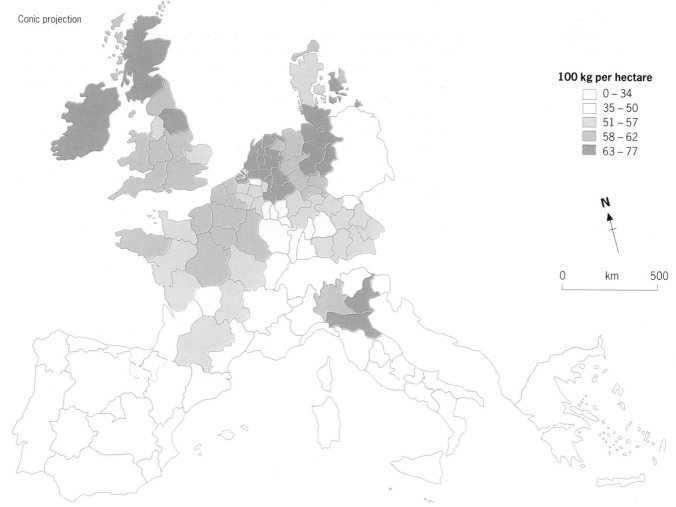

Conic projection

100 kg per hectare
- ☐ 0 – 34
- ☐ 35 – 50
- ☐ 51 – 57
- ☐ 58 – 62
- ☐ 63 – 77

N

0   km   500

Figure 8.14 Wheat yields in the EC, 1987

An assessment of von Thünen's theory

The world is not, and never has been, composed of neat spatial patterns. Von Thünen himself knew this: 'Farming systems would not succeed each other in regular succession, as in the Isolated State, but would be jumbled among each other: the farm of fertile soil, on a river, 100 miles from the Town would belong to the third ring, that 10 miles from the Town, with sandy soil, to the sixth...'

The critique of von Thünen's theory (Table 8.3) should help you reflect about the real, rather than the theoretical world of agriculture.

Table 8.3 A critique of von Thünen's theory

- The theory explains how land use *should* be organised in an idealised world, rather than how it actually is organised.

- Today we are less happy with the simplifying assumptions von Thünen made. Rather than view farmers as fully rational and perfectly informed, it is more realistic to look at the way they actually behave in a world of mixed goals and scarce information (see Sections 7.8–7.10).

- Von Thünen's isolated state was a static, unchanging place. In the real world it is often the changes which are most interesting. Thus shifting tastes in foodstuffs and increases in

consumer purchasing power have forced farmers to review their production. Meanwhile changing technologies have revolutionised transport.

- We might expect von Thünen to be most relevant in the economically developing world, where technology most resembles that of nineteenth century Europe. But there are problems with this view. Many peasant farmers produce only sufficient for their own needs and have little surplus to send to market. Often the land is not owned by individuals but collectively by the village, in which case notions of locational rent are unlikely to be relevant.

?

7 Study Figure 8.12. To what extent would you say that the distribution of market gardening supports Ricardo's theory, rather than von Thünen's?

8 Study Figures 8.13 and 8.14.

a What evidence is there of spatial patterns which support von Thünen's theory at different scales?

b What factors, other than distance to market, might influence the spatial patterns in Figure 8.14?

8.7 Marketing agricultural products

Abstract theories like von Thünen's tell us little about the complexities of modern markets. Agricultural products are very different from manufactured items and cannot be marketed in the same way. For instance, problems involved with agricultural products are as follows:

1 Perishability: most agricultural products are perishable. Thus rapid access to markets (whether the point of final consumption or a processing plant) is essential if products are to be sold in good condition and command a high price.

2 Seasonality: agricultural production is highly seasonal. This contrasts with demand which is fairly constant. The outcome can be gluts followed by periods of shortage.

3 Weather-dependence: agricultural **output** is closely tied to the weather, which may be highly variable from year to year. This presents the farmer with the difficulty of matching supply to demand.

4 Slow response to market changes: farmers face a time-lag of several months between planting a crop and selling it on the market.

We should remember that marketing is different from 'selling'. For a farmer, selling is simply disposing of crops and livestock to the highest bidder. In other words, the farmer decides what to produce and only afterwards thinks about selling it. Marketing, on the other hand, involves forward planning. The farmer aims to satisfy a known market, and uses various channels (e.g. auctions, marketing boards etc.) for selling his/her crops.

Marketing is most significant in the economically developed world. In the developing countries, most farming is still **subsistence**-based, and when surplus production occurs, this is often difficult to store and transport over long distances to markets. Even so, there have been developments in marketing in economically developing countries. The importance of plantation agriculture, and the recent expansion of fruit and vegetable cultivation for export, suggests that agriculture in developing countries is becoming more commercial and market-oriented (see Section 8.9).

8.8 Marketing channels in the UK

We shall now turn our attention to the channels through which farmers dispose of their products. In economically developed countries like the UK, the principal marketing channels are auctions, marketing boards, **co-operatives**, direct selling and contract farming.

Table 8.4 The advantages and disadvantages of livestock auctions

Advantages
• Auctions usually occur within 20 or 30 kilometres of the farm. This reduces the risk of animal weight loss on long journeys and means that the farmer does not have to leave the farm for too long.

• Everyone can see what is being sold. Livestock are transferred after the sale with minimum fuss.

• Markets have a social function, providing a meeting place and allowing an exchange of information.

• Markets are useful assembly points. Bulked cargoes can then be carried onward more efficiently.

Disadvantages
• Prices at auctions fluctuate from day to day, or even hour to hour, and it is difficult to gather this market information.

• Many auction markets are small, and prices at these markets may be lower owing to lack of competition.

• Livestock is sold 'sight as seen'. It is therefore possible to buy animals with defects that are not at first visible.

• In small markets, there is the danger of buyers forming 'buying rings' which force prices down artificially.

Figure 8.15 Hawes livestock market, North Yorkshire, UK

Livestock auctions

Auctions are a method of selling where buyers and sellers hire a professional auctioneer to organise proceedings in the market (Fig. 8.15), and the auctioneer then charges a commission to the seller. The number of auction markets in the UK is in decline, though. Those dropping out have been the smaller markets, especially the ones near to large towns and cities where farmers can sell direct to slaughterhouses and meat-packing plants. As a marketing channel, livestock auctions have both advantages and disadvantages for farmers (Table 8.4).

Marketing Boards

In the UK several Marketing Boards (e.g. for hops, milk, eggs, potatoes etc.) were set up by law in the 1930s. The Boards would buy from the farmer and sell to the consumer and so cut out the expensive middleperson (Ilbery, 1985).

The five regional Milk Marketing Boards (MMBs) were formed in 1933. One of their main functions was to sell milk to wholesalers for the liquid market, and to creameries for making butter, cheese, yoghurt and other dairy products. The MMBs were also responsible for advertising milk (along with the National Milk Publicity Council). Successful advertising has given milk a very favourable image in the UK (Fig. 8.16). Some people believe that such advertising has been partly responsible for high levels of milk consumption. Only recently has milk's positive image been threatened by scares over milk fat and people's health.

MMBs were popular with dairy farmers who received monthly payments for their milk. This gave them a steady cash flow. However, the MMBs' policy of paying farmers a standard price for their milk (within each MMB region), regardless of their location, has often been criticised. Such a policy discriminates against farmers who live at the 'pastoral core', close to the market: because of their lower transport costs, these farmers would normally have a cost advantage over more distant milk producers situated in the rural areas of the 'pastoral periphery'.

The MMBs operated a monopoly marketing scheme which subsequently met with the European Community's disapproval. This resulted in the British government abolishing the MMBs in 1994. The impact of this change on the dairy industry is likely to be considerable.

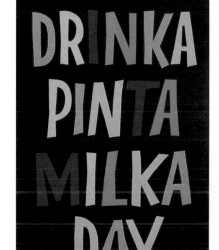

Figure 8.16 1959 advertisement for milk

9 Examine the data in Table 8.5
a How did the distribution of dairying in England and Wales change between 1924–5 and 1984–5?
b Between which dates are the changes most likely to be caused by the MMBs' pricing policy?
c What other factors might account for the changing distribution between 1924–5 and 1984–5?

10 Imagine that you were a senior official of an MMB. Write a speech for the annual conference of the National Farmers' Union, to justify your standardised pricing policy for milk. NB: You have to convince angry dairy farmers from the West Midlands and the North-West.

Table 8.5 The changing distribution of dairy farming in England and Wales (*Source*: Ilbery, 1992)

	Percentage of liquid milk sales off the farm			
	1924–5	1946–7	1954–5	1984–5
Arable England	33.5	27.7	26.5	18.1
Pastoral core	49.9	47.2	46.3	48.0
Pastoral periphery	16.6	25.1	27.2	33.9

Arable England = East, East Midlands, South, South-East
Pastoral core = West Midlands, North-West, south west Midlands
Pastoral periphery = South-West, North, Wales

Co-operatives

Farmers have a reputation for their independence, yet the history of voluntary co-operation by farmers in the UK goes back to the nineteenth century. Co-operatives have various objectives but are mainly concerned with the sale of farm produce and the purchase of inputs. The importance of co-operatives in marketing farm products is variable: for example 25 per cent of

Figure 8.17 Sussex farm advertising its own products

all vegetables in the UK are sold through co-operatives, compared to just one per cent of poultry meat. Elsewhere in the EC co-operatives have a dominant marketing position. For example, in Denmark over 90 per cent of pig meat, milk, fruit and vegetables are channelled through co-operatives.

Direct marketing

Many farmers in the UK and the EC have tried to increase their profits by selling directly to consumers. One example is doorstep delivery of milk by so-called producer-retailers. In fact, the growth of car ownership and increased personal mobility over the last 40 years has opened up new opportunities for direct marketing. Eighty-five per cent of farms in the UK have road signs advertising direct sales (Fig. 8.17). Prices are about 25 per cent above wholesale, which doubles the farmer's profit, although goods are still usually cheaper than on the retail market. Many farmers have set up farm shops and stalls. Pick-your-own (PYO) schemes, especially for soft fruit, have proved very popular in some areas (Figs 8.18–8.19).

Figure 8.18 Problems caused by picking your own!

11 Study Figure 8.19.
a Where are the greatest concentrations of PYO found?
b Why do you think that the opportunities for PYO schemes vary at:
• a local scale,
• a national scale?
c What problems do you think that PYO might create for farmers? (Look at Fig. 8.18 for some ideas)

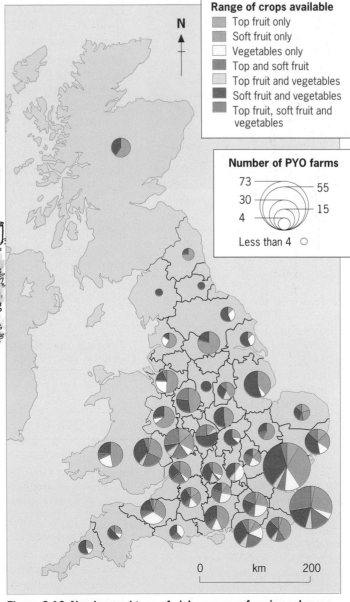

Figure 8.19 Number and type of pick-your-own farming schemes in England and Wales

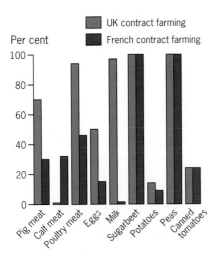

Per cent

UK contract farming
French contract farming

Figure 8.20 UK and French farm products sold in advance by contract, 1990 (Source: European Commission, 1993)

Figure 8.21 Agribusiness rotary-arm irrigation for farming on the desert fringe, Morocco

?

12 What are the advantages of contract farming to:
a the farmer,
b the food processor and supermarket chain?

13 Study Figure 8.20.
a Describe the main differences in contract farming between France and the UK.
b How does the importance of contract farming vary between products? Suggest reasons for the differences.

14 Compare and contrast agribusiness with **sustainable** farming (Section 3.10).

Table 8.6 World's largest food companies 1992 (Source: The Economist, 4 Dec. 1993)

Company	Sales ($bn)
Nestlé	37.6
Philip Morris	33.0
Unilever	24.7
Pepsi-Cola	13.7
Coca-Cola	13.3
BSN	12.4
Grand Metropolitan	9.7
RJR Nabisco	6.7
Sara Lee	6.6
CPC International	6.6
Heinz	6.6
Campbell	6.3

Contract farming
Contracts between food processors, supermarkets and farmers account for a high proportion of agricultural output in the UK. There are two types of contract. The first is where farmers are entirely responsible for production and sign a contract to deliver a given quantity and quality of a crop. Second, there are contracts which transfer management responsibility to the buyer. In the broiler chicken industry, it is common for the farmer to provide the land and labour but for the buyer to own the buildings and supply the animals. The buyer sets production targets and takes delivery of output.

8.9 Agribusiness

Agribusiness is a form of agriculture which uses modern technologies such as machinery and **agrochemicals** to produce cash crops for food and other manufacturing industries (Fig. 8.21). It is often associated with large farms and heavy capital investment.

Unlike agribusiness, traditional agriculture has never attracted great investment – mostly because of its subsistence nature. In part this reflects agriculture's poor returns and its seasonality which, in higher latitudes allows production for only a limited period of the year. This means that capital investment is not used efficiently. In contrast, multi-cropping is a feature of traditional agriculture in tropical areas. Here, conditions allow two growing seasons, three in India and in some places up to four seasons (see the Bhola Island case study in Section 7.8). Consequently, this farming system has not encouraged, or required, large-scale capital investment.

However, in recent years investors have taken a greater interest in agriculture and in the food chain as a whole. Investment into agriculture-related industries (especially agrochemicals and food processing) has been considerable. Banks have also been eager to lend money to farmers. Off-land enterprises such as pigs, poultry, and horticulture, as well as large-scale farms which have advantages of **scale economies**, have been particular targets for investment.

Transnationals
The growth of agribusiness is also tied up with the emergence of huge transnational corporations (TNCs). These have investments in every aspect of the food system, from farm production to meals in restaurants (Table 8.6).

Unilever: a transnational food corporation

Unilever worldwide

Unilever is a giant Anglo-Dutch TNC and is one of the largest industrial companies in the world. It includes some 500 companies in more than 80 countries, employs 283 000 people worldwide, and sells over 1000 brands of products. Like most TNCs, Unilever's headquarters, and the bulk of its manufacturing and research and development operations, are located in the economically developed world (Fig. 8.22). TNCs such as Unilever have considerable economic power and control a large share of world trade in food and industrial crops.

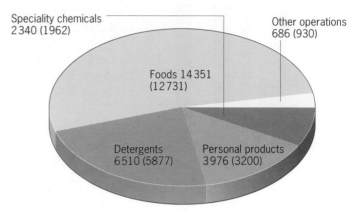

Total 27 863 (24 700)
1992 figures are in brackets

Figure 8.23 Unilever's turnover by operation (£ million), 1993 (*Source: Unilever*)

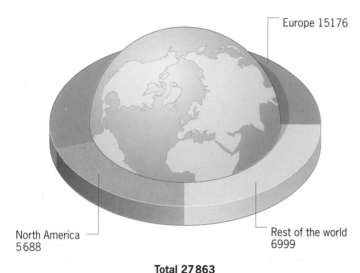

Total 27 863

Figure 8.22 Unilever's turnover by geographical area (£ million), 1993 (*Source: Unilever*)

Plantations

Unilever's interests, however, are by no means confined to the economically developed world. In 1992 the company employed 120 000 people in developing countries, where it owns and operates plantations (Fig. 8.24) and food processing, detergent and personal products factories. The plantations provide Unilever with some of the essential raw materials it needs for its various manufacturing enterprises, as well as producing for local and world markets (Fig. 8.25).

Products

Unilever's core businesses are food manufacturing, detergents, personal products – such as shampoo, cosmetics and toothpaste, and specialty chemicals – such as flavours, fragrances, starches and silicates (Fig. 8.23).

It is the world's largest producer of margarine, ice cream and packaged tea. In fact, many of Unilever's products (e.g. Flora, Oxo, Persil) and companies (e.g. Bird's Eye, Wall's, Lever Brothers) are household names.

Figure 8.24 Unilever's palm oil plantation, Malaysia

?

15a Calculate what percentage of Unilever's business is located in each of the three geographical areas (Fig. 8.22).
b Give reasons for the general pattern of Unilever's area turnover.

16a What proportion of Unilever's business is in foods (Fig. 8.23)?
b Calculate the percentage change in turnover for each product group from 1992 to 1993.

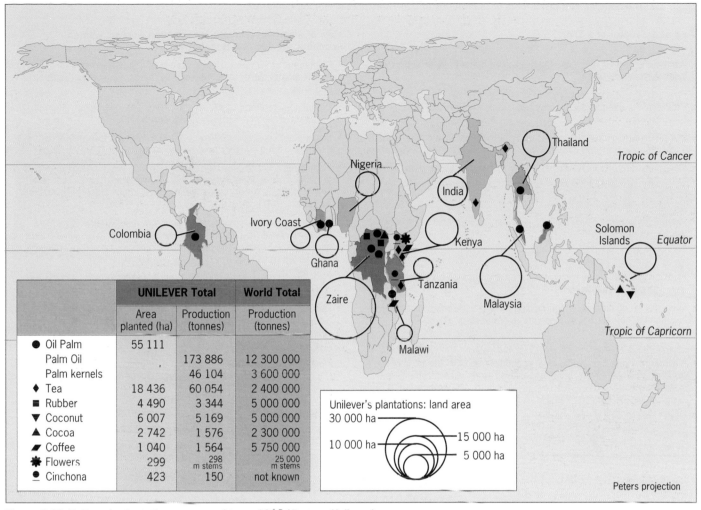

Figure 8.25 Unilever's plantations: area and type, 1993 (*Source:* Unilever)

	UNILEVER Total		World Total
	Area planted (ha)	Production (tonnes)	Production (tonnes)
● Oil Palm	55 111		
Palm Oil		173 886	12 300 000
Palm kernels		46 104	3 600 000
♦ Tea	18 436	60 054	2 400 000
■ Rubber	4 490	3 344	5 000 000
▼ Coconut	6 007	5 169	5 000 000
▲ Cocoa	2 742	1 576	2 300 000
◣ Coffee	1 040	1 564	5 750 000
✻ Flowers	299	298 m stems	25 000 m stems
☻ Cinchona	423	150	not known

Unilever's plantations: land area
30 000 ha
15 000 ha
10 000 ha
5 000 ha

17a In which parts of the world are Unilever's plantations located?
b Identify the main crops produced on these plantations.

Unilever established its first plantations in Africa during the 1900s. These produced palm oil for the company's margarine and soap factories in the Netherlands and the UK. In the post-war period, the company has favoured investment in South Asia and South-East Asia where growth prospects are higher and where there is greater political stability than in many other economically developing countries. Plantation agriculture is capital intensive and requires high levels of technical and managerial skills. Modern plantations are large (2000 to 3000 hectares) and specialise on a single crop to obtain lower costs through economies of scale.

8.10 Tropical commercial agriculture

Plantation agriculture is an example of tropical commercial farming in the developing world. It relies heavily on foreign capital, and most of the production is for export to the developed world. This contributes to the dependency of many poor tropical countries on foreign TNCs and the rich developed world. It also leads to the question of who actually benefits from plantation agriculture (Table 8.7).

Table 8.7 The benefits and disbenefits of plantations to economically developing countries

View of a director of a TNC	View of a non-government organisation working with an economically developing country
• Plantations provide much needed employment and pass on important skills to the workforce.	• Plantations are a form of 'neo-colonialism'. They are part of the system of exploitation of the developing world by the economically developed countries.
• Production is often for local as well as export markets.	• Production is geared to export and profits flow overseas to the rich countries of the economically developed world.
• Plantation agriculture often opens up remote areas. Investment in roads improves rural transport infrastructure and encourages rural development.	• Management is dominated by ex-patriate workers. Indigenous people are excluded from the higher paid/skilled jobs.
• Plantations are often established in newly cleared areas and do not take farmland from native people.	• Plantations often occupy high-quality farmland which could be used to grow food crops for the people. In the Philippines, 55 per cent of all farmland grows export crops, while calorie intake among Filipinos only averages 1940 kcal/day.
• Plantations introduce new technologies such as irrigation to rural areas.	• Plantations rely on foreign capital and expertise. Control is centred in economically developed countries, placing poor countries in a position of dependency. If profits fall, plantations might close and production be transferred elsewhere.
• Production for the home market reduces dependence on imports and saves foreign exchange.	• Plantations often rely on imported seeds, fertilisers and machinery etc., and so bring limited benefits to developing economies.
• Many plantation crops are high-yielding perennials. Compared to annual crops in the humid tropics, these crops are ecologically sound.	• Plantations encourage dependence on cash crops for export. Prices for these crops fluctuate wildly on the world market and economically developing countries have little control over world agricultural prices; their value compared to manufactured goods fell throughout the 1980s.
• Plantations inject money into local economies and this has an important regional multiplier effect (bringing positive results in employment, and status for an area). They may provide the social infrastructure for a whole community, e.g. schools, housing and hospitals.	• Plantations foster a dual economy in economically developing countries. Commercial agriculture based on wage labour may undermine traditional peasant agriculture.

?

18a Consider the beliefs of the two decision-makers in Table 8.7 and then clarify your own attitude towards plantation agriculture in the economically developing world.
b Write an article for a charity magazine (e.g. Oxfam) setting out your attitude and beliefs towards plantation agriculture in the developing world.

Summary

- In commercial agriculture efficient transport systems are vital to link farmers to markets.
- In economically developed countries, farming in the rural-urban fringe may be adversely affected by the expansion of urban areas.
- Farmers may adjust to urbanisation by changing crops and decreasing the intensity with which they cultivate their land.
- According to von Thünen's theory, the cost of transporting agricultural products to market affects spatial patterns of agricultural land use and farming intensity.
- Von Thünen's theory is abstract and highly simplified; because many factors influence land use in the real world it has proved difficult to isolate the effect of a single factor i.e. the market.
- Ricardo's theory shows how soil fertility and population density influence rents and agricultural land use.
- In commercial agriculture in the economically developed world, farm products are marketed through several channels, including auctions, marketing boards, co-operatives, direct selling and contracts.
- Agribusiness is a large scale farming operation, based on modern technology and high inputs of agrochemicals.
- The growth of agribusiness is associated with giant transnational food corporations, involved in every stage of the food system from farming to selling meals in restaurants.
- Plantation agriculture is an important commercial farming system in many economically developing countries. It has often been controlled by TNCs based in the economically developed world.

9 Food: wholesale and retail

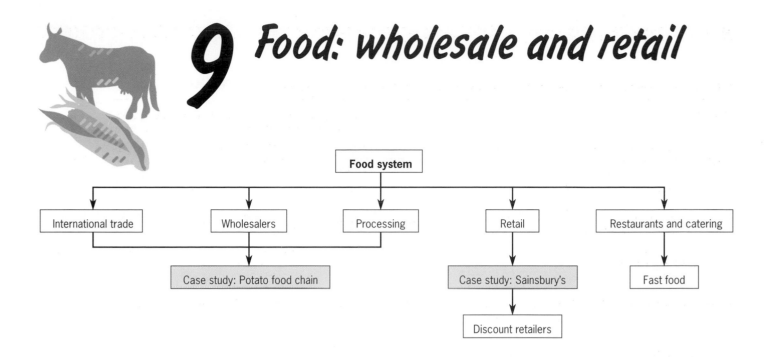

9.1 Introduction

Farming in economically developed countries is part of a complex **food system**, which increasingly is dominated by food processors, food retailers, fast-food and restaurant outlets (Fig. 9.1). The food system comprises chains of people and institutions and is the means by which urban and industrial societies obtain their food supplies. Essentially, farming is the production side of this system. Our focus in this chapter is on those parts of the food system which extend beyond the farm gate, with particular emphasis given to economically developed countries.

Figure 9.1 The UK food system (figures refer to numbers employed)

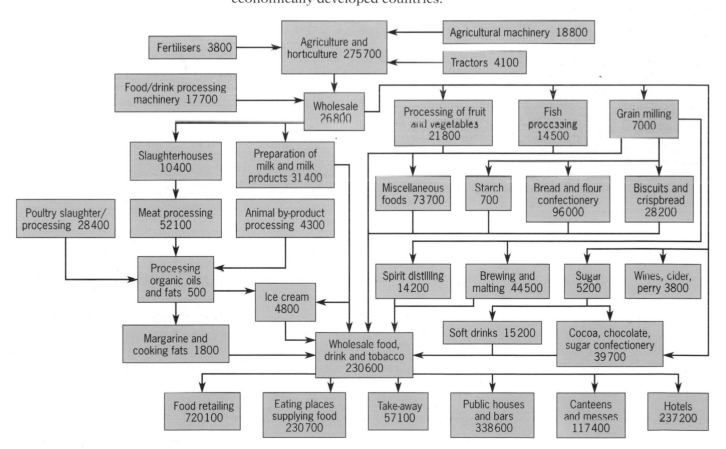

?

1 Use Figure 9.1.
a Draw a simplified flow diagram of the UK's food chain to include agriculture, trade, food wholesalers, food processing, food retailing and consumers. Make the boxes in your diagram proportional in area to the employment in each sector.
b Describe the various pathways in the food system for:
• grain milling, • sugar.

9.2 Food systems in the economically developing world

The food system concept is most relevant to economically developed countries and to **commercial agriculture** which is geared to market demand. **Subsistence agriculture**, in which farm families consume most of their own produce, has only limited relevance to food systems.

Even so, we must not forget that a sizeable proportion of food production in economically developing countries is marketed (see Section 8.7). Although most food channels may extend little further than the nearest village market (Fig. 9.2), many cash crops are grown for export (see Section 8.10).

The Gambia

In The Gambia in West Africa (Fig. 9.3), peasant farmers grow a combination of subsistence and cash crops (Fig. 9.4). Subsistence crops such as rice and sorghum are mainly consumed by farmers and their families, and so food chains are short. Groundnuts are The Gambia's principal export and, after milling in the capital Banjul, there is an international food chain comprising wholesalers, food processors and retailers in the developed world. Recently, the Gambian government has attempted to diversify agriculture by promoting horticulture. Fresh vegetable cultivation, based on irrgation and mainly female labour, is aimed at the international market. Using air freight it is possible to market fresh vegetables in the EC, especially during Europe's winter months when demand is high. In addition, Gambian farmers sell vegetables directly in their local markets. Those who farm near to the capital have the additional advantage of selling fresh vegetables to hotels which cater for foreign tourists.

9.3 Trade in food

The UK

The food system of any country or trading bloc receives food **inputs** either through the production of its own farmers or through imports. In the UK, the abolition of the Corn Laws in the 1840s was a turning point in the country's food system. A new free trade policy then encouraged the import of cheap food, especially from the British Empire and the USA. These imports, which were traded for manufactured goods, were essential to feed a rapidly expanding urban population.

By the early twentieth century, UK farmers provided only one-third of the nation's food requirements.

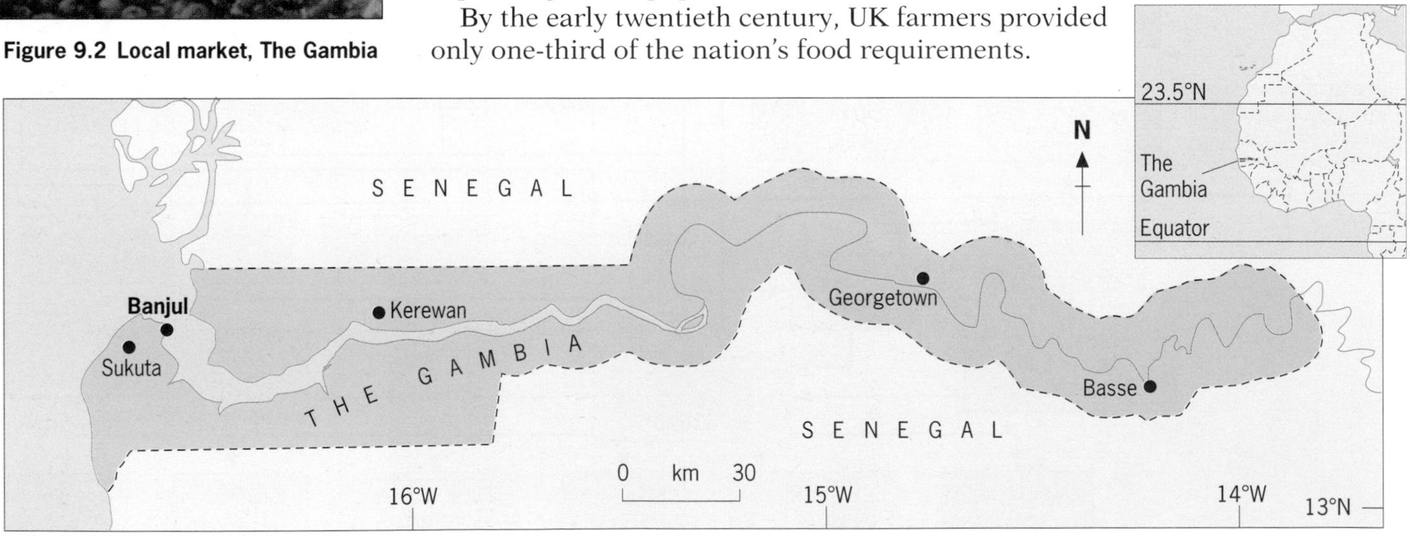

Figure 9.2 Local market, The Gambia

Figure 9.3 The Gambia

2 Use Figures 9.1 and 9.4 to identify the main differences between the food systems of the UK and The Gambia.

3a Using the data in Table 9.1, draw bar charts to show agricultural production and self-sufficiency for products in the EC.
b Explain what your graph tells you about the pattern of the EC's trade in food.
c Make a list of the factors which might account for differences between countries in their degree of self-sufficiency in various agricultural products (look back at Chapter 4 for some ideas).

4 You are a junior minister for the MAFF. You have been asked to prepare a short report for a select committee on UK self-sufficiency. Use Table 9.1, Figure 9.5 and the following questions to help you.
a In which agricultural products is the UK self-sufficient?
b Compare UK self-sufficiency with the EC. Identify the main differences, and suggest reasons for them (see Chapter 4).

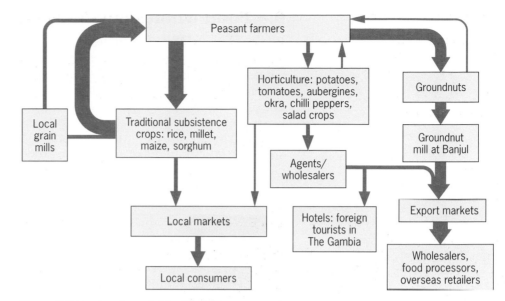

Figure 9.4 Food systems in The Gambia

However, in the second half of the century this trend was reversed. When the UK joined the EC, it was forced to adopt a more protectionist policy, with greater emphasis on self-sufficiency. Thus by 1992, the UK produced 56.5 per cent of its food and feed requirements (or 72 per cent if we exclude tropical food crops which cannot be grown in the UK)(Fig. 9.5).

The EC

The EC is a major food trading bloc. Its farm policies have succeeded in making the EC self-sufficient in many agricultural products. On the one hand, these policies subsidise EC farmers (see Section 12.4), and on the other hand, they impose tariff barriers to keep out overseas competition. Meanwhile improved crop varieties have also strengthened self-sufficiency (Table 9.1). For example, only 70 per cent of wheat used by the UK bread making industry in 1981–3 was home-grown. That proportion rose to nearly 90 per cent in 1991, thanks to the development of new high protein wheat varieties able to tolerate the UK's cool summers.

Figure 9.5 International food trade of the UK (£ million), 1991

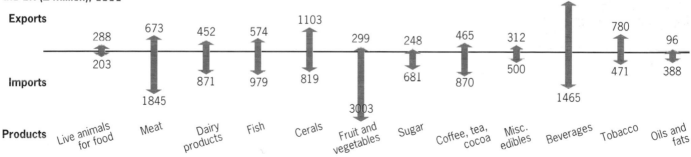

Table 9.1 Percentage of self-sufficiency in selected agricultural products, 1990–1 (*Source:* European Commission)

	Wheat	Fresh vegetables	Fresh fruit	Eggs	Beef	Pork	Poultry meat	Sheep meat	Oils: meat and fats
Denmark	178	55	20	104	209	366	220	15	99
France	238	89	86	108	117	87	137	90	82
Portugal	29	121	90	101	71	94	98	188	30
EC total	136	106	85	103	107	104	105	90	70

9.4 Food wholesalers

Wholesalers in all countries deal in bulk with farm inputs and food (Fig. 9.6). They operate at several stages in the food system (Fig. 9.1). One view of wholesalers is that they are essential to the smooth running of the food system because without them, it would be difficult to achieve a balance between supply and demand. Wholesalers try to achieve this balance by controlling prices. However, sometimes the market becomes glutted and wholesalers are left with too much stock. At other times, when there are food shortages, they can make easy profits. Indeed, this was a major factor causing the inflation of food prices in Bangladesh during the 1974 famine (see Section 10.5).

Buy produce from farmers

Buy feed and agrochemicals from manufacturers

Buy processed foods from factories

Warehousing and storage

Pricing

Grading, quality control and packaging

Distribution

Supply fresh and frozen produce to retail outlets

Supply feed and agrochemicals to farmers

Supply processed foods to retail outlets

Figure 9.6 The role of wholesalers in the food system: selling in bulk at Covent Garden Market in London, UK

?

5 Work in groups.
a Discuss why British supermarket chains are performing their own wholesaling functions.
b Suggest the possible dangers this poses, and to whom.
c Write a short article commenting on the impact, or potential impact, of supermarket wholesalers on your local area, and on the people involved. Express your own attitude to such changes.

Changing importance of wholesalers
The former Soviet Union, under its centralised **command economy**, showed how an absence of wholesalers in the food system can lead to food supply problems. Without the distribution network provided by wholesalers, crops produced on **state** and **collective** farms could not be stored, graded and transported to market efficiently in order to satisfy demand. Basic food shortages resulted, causing considerable frustration among consumers. It has been estimated that 20 per cent of Soviet food production rotted at some point on its journey from the fields to the shops because of inadequate transport, poor storage facilities, and a shortage of refrigeration for perishable crops.

In the UK's food system, wholesalers are currently losing some of their market power to the large supermarket chains. Companies such as Sainsbury's and Tesco are increasingly meeting their own wholesale needs by contracting suppliers, farmers and food manufacturers directly, and by operating their own warehouses and distribution depots.

9.5 Food processing industries

Food is one of the world's largest industries. Food (and beverage) manufacturing was worth £14.5 billion to the UK in 1991. This represented 10.6 per cent of the value of all manufacturing industries in the UK (Table 9.2). With a total employment of over half a million people, food processing was the second largest employer among **primary** and **secondary** industries in 1991 (Fig. 9.7).

Table 9.2 Production of major processed foods in the UK (thousand tonnes)
(*Source*: Central Statistical Office)

	1981	1986	1991
Flour	3596	3702	3846
Sugar from home-grown beet	1092	1318	1220
Poultry meat	747	919	1063
Beef	1053	1058	1017
Pork	710	747	777
Oil from crushed oil seeds	449	431	597
Glucose	434	472	557
Margarine and other spreads	398	460	465
Mutton and lamb	263	289	386
Cheese	242	259	298

There are two groups of food processing industries:

1 Those which process the raw output of agriculture either for later final manufacture (e.g. flour milling, sugar refining), or to preserve produce in a near-fresh form (e.g. bottling, canning, and freezing fruit and vegetables).

2 Those industries which manufacture branded goods for the consumer (e.g. soup, biscuits, breakfast cereal etc.).

The UK has strong food processing industries which compete effectively in world markets. For instance in 1983, of the 100 largest food corporations in the world, 22 were based in the UK. The UK also had 15 of the top 21 European food manufacturing companies (by sales).

There is a problem, though, for food manufacturers in that as people become better off, the proportion of their income spent on food begins to fall. For example, in the USA 16.2 per cent of disposable income was spent on food in 1970, compared to 13.5 per cent in 1991. To maintain their profits, manufacturers have moved to products with higher value-added, such as processed and convenience foods. Value-added results when the retail cost of items are increased because of extra production costs. Using advertising to shape demand, manufacturers have also successfully promoted products such as ready-made meals cooked in microwave ovens, quick snacks, health foods, light foods and drinks for dieters. This has only been possible through the development of new flavourings and preservatives, quick freezing, dehydration, and computer control.

Figure 9.7 The food processing industry: checking the quality of potato crisps

9.6 Modern food systems

In this section we shall investigate how a modern food system works by taking the example of potatoes in the UK.

The potato process in the UK

Potatoes have been a standard part of the British diet for at least 200 years (Fig. 9.8). They have proved extremely popular as a cheap source of carbohydrates and essential nutrients such as vitamin C. However, in recent decades, as incomes have risen for most people, the humble potato has appeared less attractive. This threat to potato consumption presented the food processing industry with a considerable marketing challenge. They responded by introducing a wide range of potato products. Crisps of many flavours have been joined on the supermarket shelves by frozen chips, filled jacket potatoes, and a variety of potato-based savoury snacks.

Potato growing

Although it is possible to grow potatoes in most parts of the UK, the distribution shows strong regional concentrations (Fig. 9.11). This concentration is partly due to physical factors and partly to the location of potato processing plants in a belt of counties from Suffolk to Humberside in England, and from Fife to Tayside in Scotland (Ilbery, 1992).

?

6   Study Figure 9.8.
a   What proportion of potatoes in the UK are imported?
b   What proportion of potatoes in the food chain are processed?

7a   From Figures 9.9 and 9.10, describe potato sales in the UK between 1986 and 1992.
b   Which potato product has the biggest single market? Suggest what this market is.

8   Study Figure 9.11.
a   Describe the distribution of potato growing in the UK.
b   Refer back to Chapter 4 and explain how physical factors might influence the distribution of potato growing in the UK.

The potato harvest takes place over several months and is divided into 'earlies' and 'maincrop'. 'Earlies' grow quickly and are ready to harvest in early summer, when they fetch higher prices in the shops. 'Maincrop' potatoes, which are more widely grown, are harvested between late August and early November. These may be stored and then sold until May of the following year.

Potato marketing

Despite vigorous sales of a wide range of potato products, potatoes have not featured at all in the EC's Common Agricultural Policy (CAP). In the UK, though, potato production has been strictly controlled by a quota scheme administered by the Potato Marketing Board (PMB). This Potato Marketing Scheme limits the area of potatoes grown and gives the PMB some control over production levels. In addition, the PMB guarantees to purchase any grower's surplus – although at a reduced rate. By 1997, the intervention scheme will cease, in response to government attempts to introduce a free market.

Potato processing

Between 1987 and 1993 the proportion of potatoes used for processing in the UK rose from 1442 to 1806 thousand tonnes. In 1990, food processing companies and supermarket chains contracted directly with growers to take 14 per cent of the UK's potato crop. Contract marketing is an important trend which has had a major impact on potato farming. Growers have

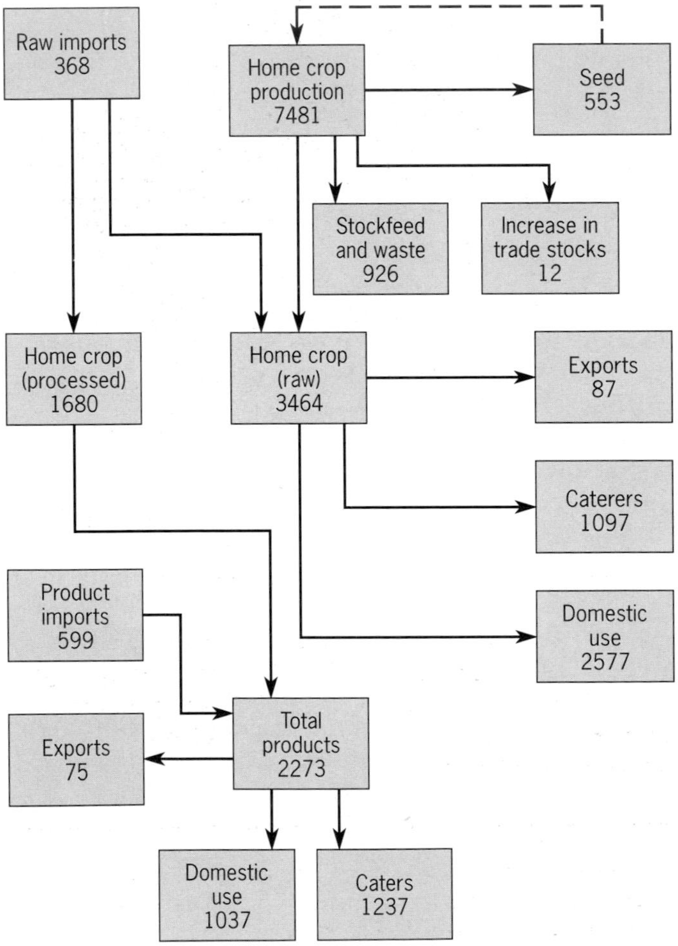

Figure 9.8 Potato use in the UK, 1991 in thousand tonnes (*Source:* MAFF, Potato Marketing Board)

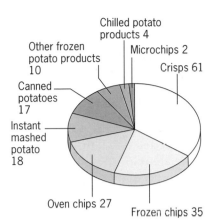

Figure 9.9 **Purchases of potato products in the UK, 1986-7 (*Source*: Potato Marketing Board)**

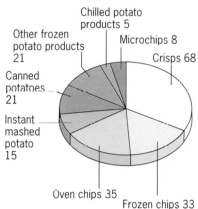

Figure 9.10 **Purchases of potato products in the UK, 1991-2 (*Source*: Potato Marketing Board)**

Figure 9.11 **Distribution of potato-growing in the UK**

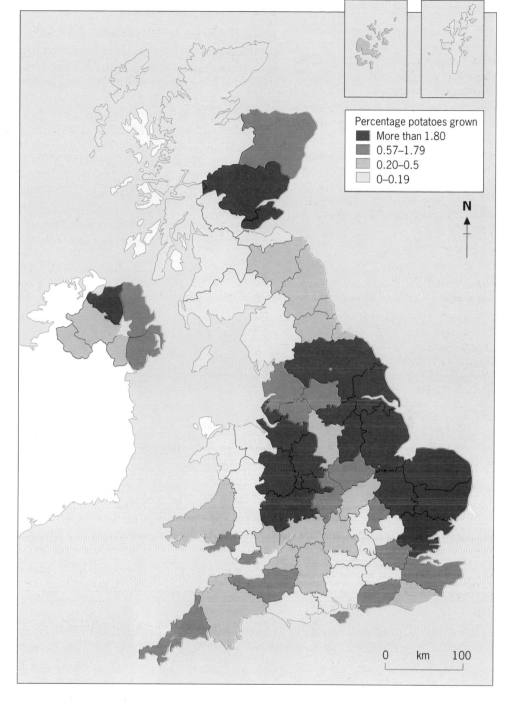

to produce higher quality and more standardised crops of a specific size, shape and texture. In fact, contract farming and the close relationship between growers, food processing industries, and food retailers has considerably lengthened the potato food chain.

Varieties

It is important to appreciate that there are many different varieties of potato. Each one has its own qualities and is best for a particular kind of use. For instance, the McDonald's fast-food chain only uses the Russet Burbank variety which has the right dimensions and texture for their French fries. A single processing plant may use several different potato varieties to make 'own-brand' products for individual supermarket chains.

?

9 Consider the impact of contract marketing by drawing a matrix to show its advantages and disadvantages for: • the grower, • the retailer, • the consumer.

10 The major grocery chains account for 61 per cent of employment in food retailing in the UK (Fig. 9.12), but generate 76 per cent of turnover (Fig. 9.13). Suggest reasons for the difference between employment and turnover.

11a List the differences between the methods of shopping in Figures 9.14 and 9.15.
b Compare each type, stating which method you prefer and why.

9.7 Food retailing

Towards the end of the food chain we find food retailers, restaurant and catering establishments. It is here that food finally reaches the consumer. Europeans spend a staggering US$340 billion on food each year. It is therefore hardly surprising that food retailers account for most employment and value-added in the UK food system (Figs 9.12 and 9.13).

Food retailing has undergone a revolution in economically developed countries since the 1960s. The dominance of traditional open markets and independent shopkeepers (Fig. 9.14) has been superceded by giant food companies. These have gained a steadily increasing share of the retail market since World War 2 (Fig. 9.15). In the UK, three big supermarket chains – Sainsbury's, Tesco and Safeway – control one-third of the country's grocery market. In 1965, food superstores were unheard of; in 1993, there were nearly a thousand. In 1980, only 12 per cent of grocery sales were on edge of town sites; in 1993, the proportion was 50 per cent.

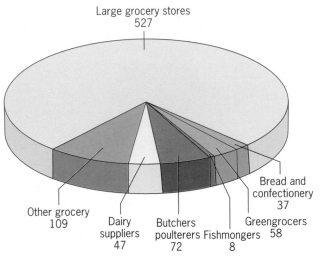

Large grocery stores
527

Other grocery
109

Dairy suppliers
47

Butchers poulterers
72

Fishmongers
8

Greengrocers
58

Bread and confectionery
37

Figure 9.12 Food retail employment (thousands) in the UK, 1990 (*Source:* Central Statistical Office)

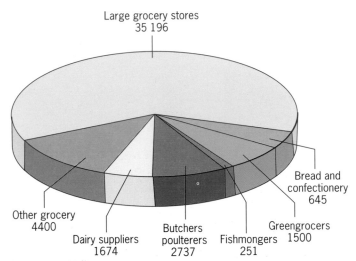

Large grocery stores
35 196

Other grocery
4400

Dairy suppliers
1674

Butchers poulterers
2737

Fishmongers
251

Greengrocers
1500

Bread and confectionery
645

Figure 9.13 Food retail turnover (£ million) in the UK, 1990 (*Source:* Central Statistical Office)

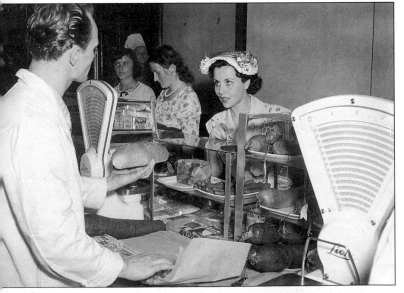

Figure 9.14 Personal over-the-counter assistance during the 1950s, Germany

Figure 9.15 Supermarket self-service in 1994, UK

Sainsbury's: a giant food retailer in the UK

Sainsbury's is the UK's largest grocery chain. Its first shop opened in Drury Lane in London in 1869. By 1900 the company had 48 outlets, all in the South-East, which sold groceries across the counter. The first self-service branch was converted in 1950, and the transition to supermarket retailing was complete by 1982. The development of edge of town sites, geared to car-borne shoppers, began in 1972. In the mid 1990s two-thirds of Sainsbury's customers arrived by car.

An outstanding trend in food retailing over the last 20 years has been the increasing size of stores (Fig. 9.16). In 1993, Sainsbury's new stores ranged in floorspace size from 3250–4500 m² and sold up to 17000 different lines. At the same time, Sainsbury's operated 328 supermarkets, and by mid 1994 the company employed 93500 staff, and served 7.5 million customers a week.

The size and power of companies like Sainsbury's in food retailing means that they can insist on high standards from their suppliers. In effect, they demand the right quality and quantity of produce at a time that suits their own retailing schedule. Many farmers find it difficult to meet these rigorous standards. However, while supermarket companies like Sainsbury's are capable of influencing both growers and public opinion, in recent years they have also responded to market demand: since the early 1980s Sainsbury's has become more environmentally conscious. Examples of this include packaging minimalisation (which lowers the cost of both packaging and transport) and integrated crop management schemes (which reduce production costs). Growers have been encouraged to reduce their use of pesticides, and to place more emphasis on the biological rather than the chemical control of pests. In 1991, a scheme was introduced to supply beef and lamb reared by traditional methods. This now has 1500 farmer members and has been extended to include pork from pigs reared outdoors.

The distribution of Sainsbury's stores

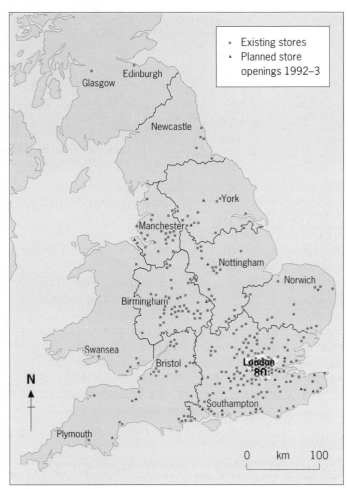

Figure 9.17 Distribution of Sainsbury's stores, 1991 (*Source:* J. Sainsbury plc)

12 Study Figure 9.17.
a Test the hypothesis that diffusion from London and the South-East explains the distribution of Sainsbury's stores. Method: count the number of stores in six 100 kilometre zones around London; use Spearman's rank correlation test (Appendix A1) to find the association between distance from London and number of stores; test for significance in the t-tables and comment on your results.
b Test the alternative hypothesis that Sainsbury's stores are market-oriented. Method: count the number of stores in the planning regions shown in Figure 9.17; correlate the population of each planning region (Table 9.3) with the number of stores in each region; test for significance and comment on your results.
c Using evidence from your hypothesis testing, describe and explain the distribution of Sainsbury's stores in the UK. Apart from diffusion and market potential what other factors might explain the distribution pattern?

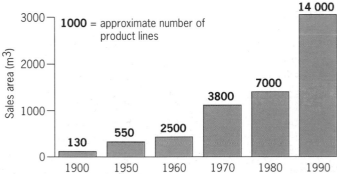

Figure 9.16 Sainsbury's stores average retail floorspace (*Source:* J. Sainsbury plc)

Table 9.3 UK Standard planning regions

Region	Population (million)
North	3.08
West Midlands	5.15
East Midlands	3.78
East Anglia	1.88
South-East	16.89
South-West	4.34
North-West	6.45
Wales	2.77
Scotland	5.15
Yorks and Humberside	4.88

Figure 9.18 Catering for the car-borne shopper; an edge of town Sainsbury superstore

The distribution of Sainsbury's stores reflects the company's origins in South-East England. Until the 1960s, Sainsbury's was reluctant to extend its operations beyond the daily delivery range of its headquarters in central London. In 1994 there were still 80 shops within the ring of the M25, and the centre of gravity clearly remains in the South-East. Since the 1970s, however, Sainsbury's has rapidly expanded into the Midlands and North-West. In comparison, several other major supermarket chains have distributions which still reflect their company's region of origin. Examples include Kwik Save in North Wales, Tesco in the South-East, and Morrison's in West Yorkshire. They illustrate the concept of spatial diffusion that we examined in Chapter 7.

Delivery systems

As Sainsbury's became a national grocery chain, it had to improve its delivery systems. Sainsbury's have four distribution depots. These are regional distribution centres which supply the full range of goods sold by Sainsbury's supermarkets. Deliveries from these to stores are made by huge 38 tonne trucks (compared to 7.5 tonnes in the early 1960s), which reduces transport costs. Most deliveries are made at night.

Storage and supply policies for different goods

The handling needs of each type of product sold in Sainsbury's are all different. In fact, their storage and supply varies as much as their consumption patterns. The supermarkets themselves have limited storage space. This on-site storage, or warehouse area located behind the shop, is used mainly to keep stocks of non-perishable, fast-selling lines such as large cans of baked beans. These warehouse lines are replenished daily from the central depot.

Other non-perishable products are only held on each supermarket's shelves – either because they have a relatively low 'stock turn', or because enough of the product can be displayed on shelves to meet shoppers' daily needs. As with the warehouse lines, these are then re-stocked at night.

The head office computer holds information about each supermarket's requirements and these are ordered daily or weekly from the regional supply centre. Because of this efficient stock control, using data collected by point of sale computers, levels of stock held 'behind the scenes' are reduced. In fact, Sainsbury's warehouses stock no more than ten days' worth of non-perishable goods.

Perishable items risk deterioration. As a result, only a few extra stocks are kept on site in the warehouse and new supplies are monitored by the central computer. The accuracy of sales information and forecasting through point of sale data means that orders are made on projected sales. Production cycles are therefore much more closely geared to demand and Sainsbury's and suppliers work very closely together.

Supermarket layout

Sainsbury's supermarkets have a standard format where the main rectangular sales area is backed by the food preparation and warehousing area. Their edge of town supermarkets cater for shoppers arriving by car and so provide extensive parking space (Fig. 9.18). Most stores carry between 8000 and 17 000 different lines. Figure 9.19 shows the internal geography of a typical supermarket. The layout has a logic and is planned not only to be convenient and attractive to shoppers, but also to maximise potential sales.

?

13 Suggest how the layout of a supermarket (Fig. 9.19) reflects the retailer's aim of maximising sales. For example, why are fruit, vegetables and flowers often located close to the entrance; chilled foods concentrated in one area; and alcoholic drinks often furthest from the store entrance?

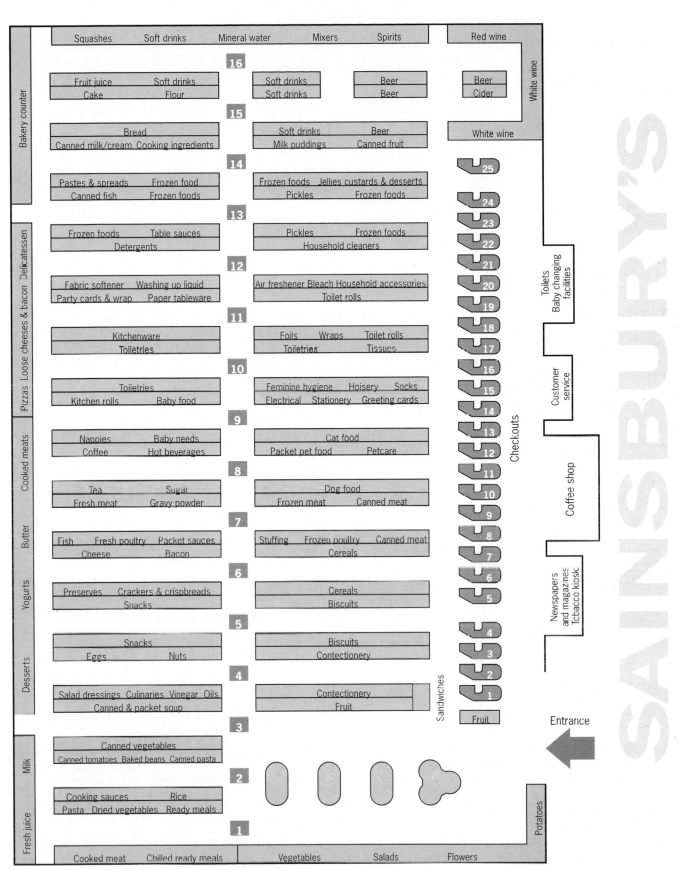

Figure 9.19 Internal layout of a Sainsbury store (*Source:* J. Sainsbury plc)

?

14 Read the article in Figure 9.20.
a Explain how the location policies of supermarket chains like Sainsbury's has contributed to the development of a 'two-tier' diet in the UK.

b Explain how the growth and location of discount stores in the UK has contributed to a 'polarisation of diet and shopping patterns'.

Figure 9.20 When big food stores move out of town, what price the consumer's health? (*Source: The Independent on Sunday*, 23 May 1993)

9.8 Supermarket chains under threat: the rise of discount retailers

The major supermarket chains in the UK face increasing competition from discount retailers such as Kwik Save, Netto and Aldi. Discount food stores are able to undercut substantially the prices at the major supermarket chains – sometimes by as much as 40 per cent. They achieve this by reducing their overheads and operating costs; concentrating on a narrow range of fast-selling products, including very few perishable items (which demand more complex handling and storage); using basic display techniques; providing only limited facilities such as car parks and toilets; and working on profit margins as low as one per cent (compared to an average of six per cent for supermarket chains). In Germany, discounters already control 25 per cent of the retail food market.

In 1993, a further shopping outlet appeared: the first warehouse club in the UK (Costco) received planning permission to build a store at Thurrock in Essex. Costco already has one hundred outlets in the USA. For a small annual membership fee it sells a wide range of products (including food) to

Rambutan for the rich, beans for the broke

David Nicholson-Lord reports

The fresh fruit counter at a new superstore is packed full of vitamin C from exotic sources: rambutans from Thailand, physalis from Colombia. For middle-class dinner parties, they made a good ice-breaker. For millions of others, the message is less wholesome. According to EC criteria, there are around 12 million people in poverty in Britain: their disposable income is less than half the average. During the 1980s, while the nation as a whole was eating more fibre, fruit and vegetables, thousands of poorer families were seeing their diets worsen. Some larger households in 1991 were eating only a third of the fresh green vegetables they consumed in 1980; their meat and fresh fruit consumption halved.

Many health specialists believe we are seeing the emergence of a 'two-tier' diet in Britain – one for the well-off, another for the poor. The implications for health are enormous. Unemployment, coupled with taxation policies and cuts in the real value of social security benefits, is partly to blame. But there is another, less obvious culprit – the supermarket wars.

In the pursuit of larger floor areas, better-off customers and bigger margins, firms such as Sainsbury's and Tesco have been deserting the city centres for leafier suburban and edge-of-town sites. In 1988, for example, 65 per cent of new superstores were opened on the edge of towns but only 10 per cent in shopping centres. But what about the people left behind? According to Spencer Henson, a retail economist at Reading University, the

costs have been borne by consumers, in terms of time and transport, but they have been 'prohibitively high' for people without the use of a car, the poor and the physically disabled. 'Low-income consumers, who in any case spend a higher proportion of their income on food, are often unable to reach the food shops with some of the lowest prices in the UK.'

This ought to be good news for the corner shop. But it often no longer exists. Since 1962 the number of independent grocers has fallen from 116 000 to 32 800.

According to Dr Henson, the flight of the superstores created a 'vacuum' in town and city centres. Increasingly it has been filled by the discount store. The rise of the 'pile it high, sell it cheap' shop is a phenomenon of recession that promises to outlast it. KWIK SAVE, already the third biggest retailer of packaged groceries in Britain, is opening a store a week. European groups such as Aldi, Netto and Carrefour are being followed to the UK by Costco, an American warehouse club, which plans to open its first sites this year. Retail analysts point to the rise in discounting as evidence of the 'polarisation' of shopping patterns, between those with cars, jobs and choice, and the bus-borne bargain-hunters. It also means a more polarised diet. A Sainsbury's superstore might offer 15 000 products; KWIK SAVE sells 3 000 and some of the other discounters 600. Healthy foods are squeezed out in favour of tins and biscuits.

Poorer families want to eat more healthily, but cannot afford it. According to a National Children's Home survey in 1991, 60 per cent

of parents would buy more fruit if they had an extra £10 to spend on food for their child; 54 per cent would buy more lean fresh meat and 38 per cent more vegetables; fewer than 10 per cent mentioned cakes, biscuits and snacks.

Yet since 1974, disparities in diet between the top and bottom social classes have increased. The richest fifth consumes 20 per cent more fresh green vegetables, 70 per cent more fruit and 400 per cent more fruit juice than the poorest fifth.

'Research repeatedly shows that food expenditure, being often the only flexible item in the household's budget, is squeezed when times are hard.' Because of the links between, for example, fruit, vegetables and stomach cancer, poor consumers run 'a significantly higher health risk'. One study estimated that low-income groups could suffer a 20 per cent higher death rate solely on the basis of diet.

Some believe the exodus of the supermarkets and the arrival of the discount stores reinforce such trends. 'Shops such as KWIK SAVE are very sensible about locating in poorer communities but they often carry very little in the way of healthy food. We already have a two-tier diet in Britain. This is exacerbating it.' KWIK SAVE says it includes fruit and vegetables in its stores where space permits: 500 of its 780 stores have greengrocers' concessions and the number is increasing as bigger shops are opened. It also has annual healthy eating promotions. If a product is popular, it stays.

15 Study the market penetration of Sainsbury's and Netto's supermarkets in Figures 9.21 and 9.22. We shall assume that trade areas correspond to postal sectors where there is 50 per cent or more market penetration.

a Suggest three possible reasons why the trade area of Sainsbury's supermarket 1 is more extensive than supermarket 2 (Fig. 9.21).

b What is the maximum market penetration of stores 1, 2 and 3 in Figure 9.21? What factors might explain the differences?

c Describe and account for the main differences in trade area size and market penetration between Sainsbury's and Netto's stores (Fig. 9.22). (NB: exclude the Otley Netto, which has a rural trade area of low population density.)

customers at wholesale prices. This style of warehouse trading depends on passing the cost of storage of bulk buys to the consumer. Experience in the USA shows that they therefore have limited popularity and so the established supermarket chains, such as Sainsbury's and Tesco do not consider warehouse clubs a major threat. None the less, the supermarket chains have responded to the warehouse presence by price cutting. Some supermarket chains, such as Asda and Gateway, have gone further and established their own discount formats and low-cost, own-brand goods.

Supermarket trade areas

Supermarket chains locate their stores in order to maximise sales potential. To do this they must look carefully at potential locations for a store relative to the surrounding catchment area. This is the trade area from which the

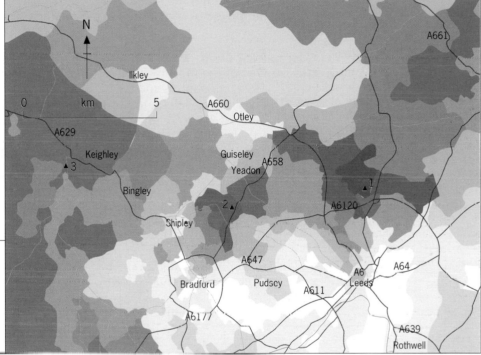

Figure 9.21 Trade areas: market penetration (by postal sector) of Sainsbury's supermarkets, West Yorkshire (*Source:* CCN Group 1992, Post Office 1992, Automobile Association 1992)

Figure 9.22 Trade areas: market penetration (by postal sector) of Netto's supermarkets, West Yorkshire (*Source:* CCN Group 1992, Post Office 1992, Automobile Association 1992)

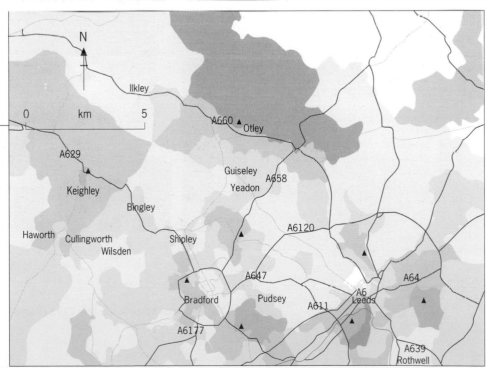

16　You are a location analyst helping Sainsbury's find a location for a new supermarket in West Yorkshire.

a　Where on Figure 9.21 would you choose to locate the new supermarket?

b　What factors influenced your choice?

17　Assume you are a planning officer. Your task is to advise your town's planning committee on proposals to build either:

• a large food superstore, or

• a smaller food discount store.

The site which has been earmarked is on the edge of town, on former industrial land. Nearby are large local authority and owner-occupied residential areas, and a small suburban shopping centre with a dozen shops, mainly selling convenience goods. The nearest large supermarket is five kilometres away. Write a report for the planning committee assessing the likely impact of the two proposals on the physical, economic and social environments of the area, clearly stating your recommendations.

store will attract its customers. For instance locational analysts will be interested in the economic, social status and spending power of the local population; their mobility; estimated journey times to the store; and the extent of competition from other supermarket chains. In addition, they will have to find a suitable site and persuade local planners to support their proposal.

9.9 Restaurants and catering

In addition to food retailers, restaurants, hotels, public houses and other catering establishments are found near the end of the food supply chain. Their combined turnover in the UK in 1990 was nearly £31 billion and they employed over one million people.

Fast food

One of the most rapid areas of growth has been fast-food restaurants. In 1990, consumers spent £4.2 billion in fast-food outlets in the UK. They bought a variety of foods, from old favourites like fish and chips and Chinese take-aways, to relative newcomers to the UK diet such as 'Big Macs' and kebabs.

In the UK and worldwide, McDonald's, the transnational food corporation, is the market leader (Fig. 9.23). It opens between 900 and 1200 new restaurants throughout the world each year, and claims to feed 0.5 per cent of the world's population every day! McDonald's first UK restaurant was in Woolwich in south-east London in 1973. By 1994, it had 526 outlets in the UK, employing 32 000 people; and by the year 2004 it aims to double its British outlets to over one thousand.

Initially located in the high street, McDonald's restaurants spread into the suburbs in the 1980s. Many of these were drive-through outlets and located in retail parks or on free-standing sites along busy routeways (Fig. 9.24). During the 1990s, the company became increasingly flexible in its choice of locations and we may yet see its fast-food restaurants in hospitals, supermarkets, cross-channel ferries and at motorway service stations.

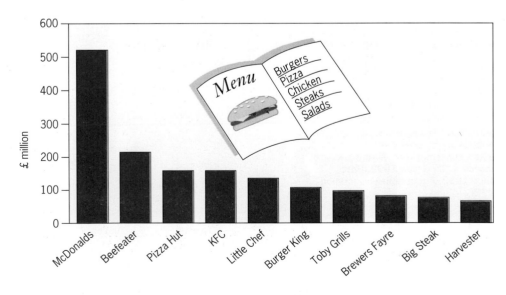

Figure 9.23 Fast-food sales in the UK, 1993 (*Source*: Technomic Consultants and SG Warburg)

18 Essay: Assess the relative importance of wholesalers, retailers and restaurants in the food system.

Figure 9.24 McDonald's restaurant, Preston

Summary

- A food system is a chain of people and institutions, starting with farmers and finishing with retailers, through which consumers obtain their food supplies.

- In commercial agriculture, food systems comprise long and complicated food chains: in peasant agriculture, food chains may not extend beyond the nearest village market.

- Some food chains in economically developing countries which include cash crops for export are as long and complex as those found in economically developed countries.

- Food enters the food chain of a country either through the production of its own farmers or through international trade.

- Food wholesalers deal with food supplies in bulk and ensure a balance between supply and demand. Currently, in many economically developed countries, wholesalers are losing some of their power to large supermarket chains.

- Food processing is one of the world's largest manufacturing industries.

- Food retailing accounts for most employment and value-added in the food systems of economically developed countries.

- Food retailing in economically developed countries has undergone a revolution since the 1960s. The importance of open markets and small independent food shops has declined, as big supermarket chains dominate the grocery market.

- The store distributions of leading supermarket chains in the UK show evidence of spatial diffusion.

- In the UK, the location policies of the leading supermarket chains and the rise of food discount stores have contributed to a polarisation of diet and shopping.

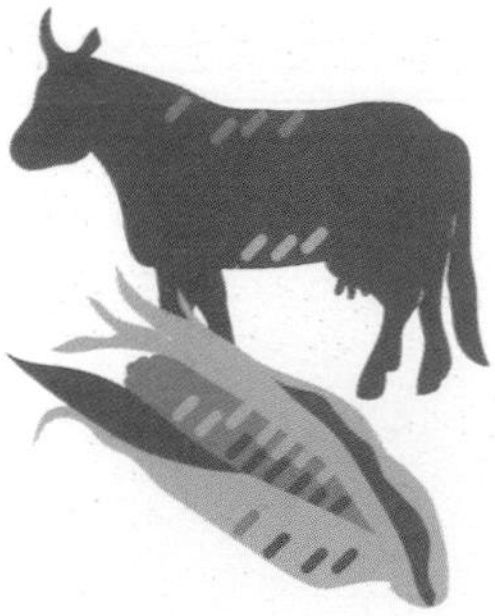

10 Nutrition, disease and famine

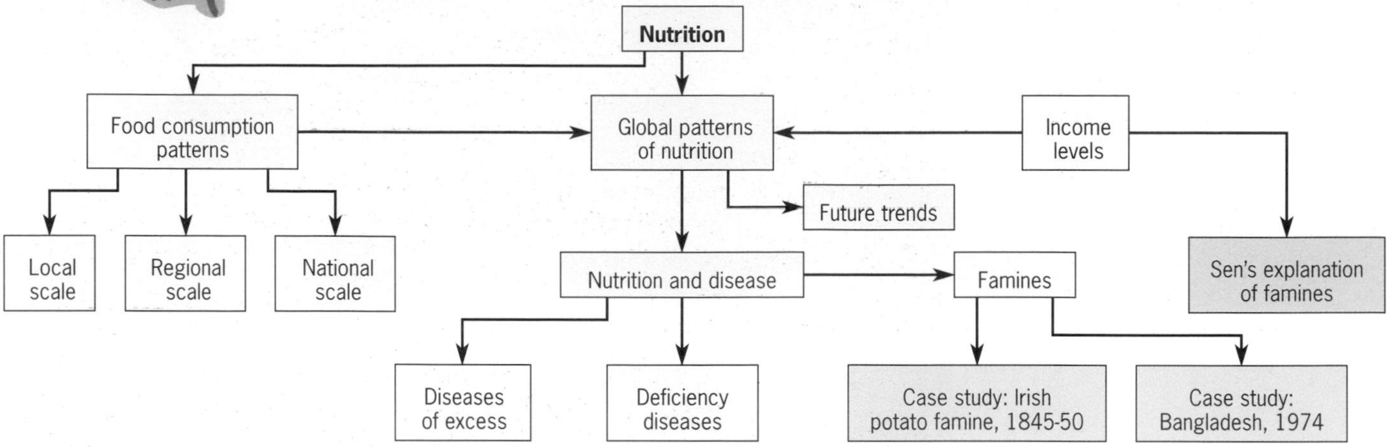

Nutrition
Food consumption patterns
Global patterns of nutrition
Income levels
Future trends
Local scale
Regional scale
National scale
Nutrition and disease
Famines
Sen's explanation of famines
Diseases of excess
Deficiency diseases
Case study: Irish potato famine, 1845-50
Case study: Bangladesh, 1974

10.1 Introduction

Everyone needs a varied diet to stay healthy. However, the geography of food consumption – what people eat, where and why – depends as much on cultural preference as on income (Figs 10.1–10.2). At the global and national scales there is considerable unevenness in spatial patterns of food consumption. On the whole, in the rich, economically developed countries, overeating is a problem. This contrasts with the problems of undernutrition, food production and distribution in economically developing countries. These are some of the most important and unresolved issues of our time. Hunger affects up to 800 million people and presents an enormous challenge not only to the countries most affected, but also to the international community.

Seasonal food shortages are common in economically developing countries, especially in the months leading up to harvest, when stocks are low and prices are high. Poor people have developed ways of coping with seasonal food shortages, and in most years they are able to get by. However,

Figure 10.1 The geography of food consumption

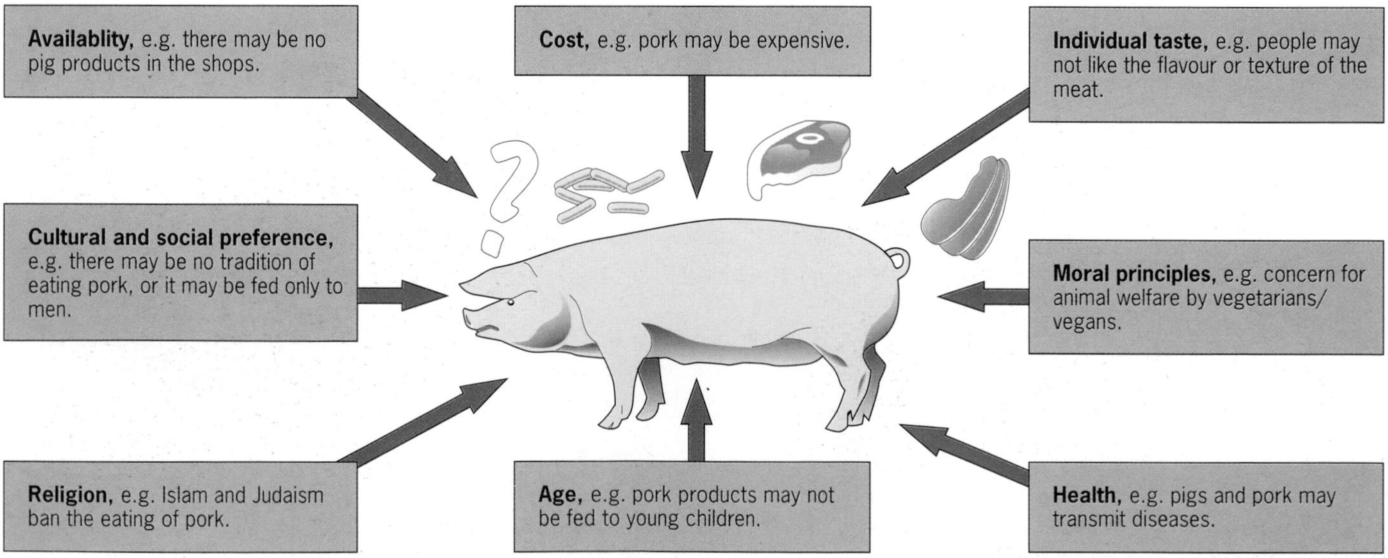

Availablity, e.g. there may be no pig products in the shops.
Cost, e.g. pork may be expensive.
Individual taste, e.g. people may not like the flavour or texture of the meat.
Cultural and social preference, e.g. there may be no tradition of eating pork, or it may be fed only to men.
Moral principles, e.g. concern for animal welfare by vegetarians/ vegans.
Religion, e.g. Islam and Judaism ban the eating of pork.
Age, e.g. pork products may not be fed to young children.
Health, e.g. pigs and pork may transmit diseases.

1a Design a questionnaire to investigate people's food preferences.
b Find out which types of food they avoid and the reasons for their preference, e.g. religious, moral, health etc.
c Do preferences vary with age, gender, social, cultural and ethnic groups etc?
d Summarise people's attitudes to food, e.g. which foods are avoided, what percentage show strong food preferences and the reasons for these preferences etc.

Figure 10.2
Cultural and religious influences give cattle a special status in India

in times of environmental, economic or political crisis they may be overwhelmed. It is then that severe food scarcity and famine become serious threats.

10.2 Patterns of food consumption

Patterns showing what people eat are evident at a range of scales, from local to global.

Local scale

A French scholar, Claude Thouvenot, conducted a survey of children's food habits in Alsace-Lorraine, close to the Franco-German border. Because Alsace-Lorraine has been disputed territory between France and Germany for hundreds of years, it has a mixture of language and cultural traditions, including diet. Thouvenot expected to find some regional patterns but was surprised at the sharpness of the divide between French and German speaking groups. Red cabbage, for instance, is a favourite delicacy of both groups, but is served differently. German speakers eat it as a cooked vegetable (Fig. 10.3): French speakers prefer it cold, as a salad vegetable (Fig. 10.4). This cultural difference reflects political divides which go back two thousand years to the time when the region marked the frontier between the Roman empire and unconquered Teutonic tribes.

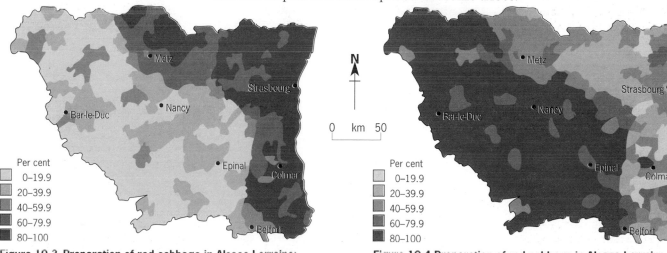

Figure 10.3 Preparation of red cabbage in Alsace Lorraine: percentage cooked (*Source:* C. Thouvenot, 1987)

Figure 10.4 Preparation of red cabbage in Alsace Lorraine: percentage used in salads (*Source:* C. Thouvenot, 1987)

Figure 10.5 Independent television regions in the UK (1989)

Regional scale

The food industry is well aware of regional differences in taste which feature in its market research and advertising of new products. The 'sour diet' of the English Midlands is an interesting example. Midlanders appear to have a particular taste for pickling (pickled onions, pickled cabbage and pickled walnuts), the acidity of which is twice what Londoners will buy. Vinegar consumption is also high in the Midlands, while Midland beer is more bitter than elsewhere. No satisfactory explanation has so far been given for this. Food consumption patterns are also evident at larger national and international scales.

Table 10.1 Regional variations in food choice in the UK, 1986–9 (*After*: Market Assessment Publications Ltd)

TV region	Mustard	Morning goods (e.g. rolls, baps, croissants, scones, toasted products)	Sponge puddings	Relish (e.g. Worcestershire Sauce, Gentlemen's Relish)
London	2	3	3	3
TVS	2	2	2	2
TSW	3	2	3	2
HTV	2	3	1	3
Anglia	1	2	3	3
Central	1	2	1	2
Yorkshire	3	2	2	1
Tyne Tees	3	2	2	1
Granada	3	2	2	1
STV/Border	3	1	2	1
Grampian	3	1	1	3

1 = <20 per cent above national average
2 = average
3 = <20 per cent below national average

?

2a Make a copy of the outline map of television regions (Fig. 10.5) and then draw a choropleth map to show the distribution of one of the food products in Table 10.1.
b Comment on the distribution. Does it show a clear spatial pattern or is the pattern random? If you can identify a non-random pattern, what factors might explain it?

3 Study the food roses in Figure 10.6.
a What are the major differences in the diets of each of the six countries in Figure 10.6 compared to the world average?
b Which country has:
• the most specialised diets,
• the most obvious food avoidances?
c To which countries would you market hamburgers? Give reasons for your choices.

Global scale

Although the consumption patterns of particular foods are of interest to food manufacturing companies and other specialists, geographers tend to concentrate on global patterns of nutrition (Fig. 10.7).

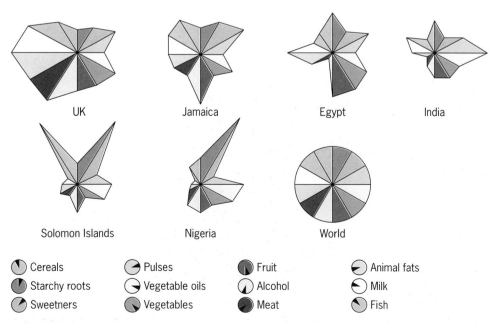

Figure 10.6 Food consumption in selected countries

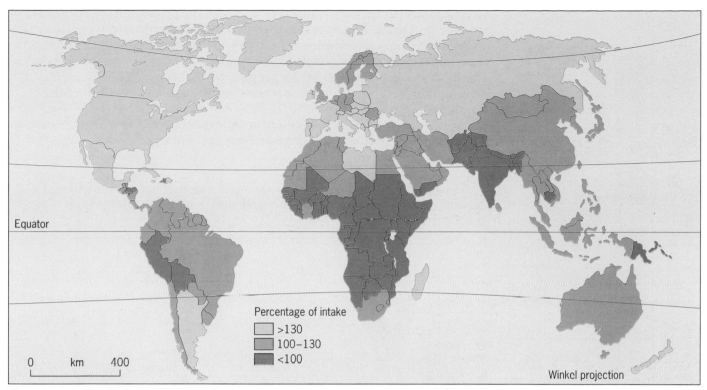

Figure 10.7 Daily calorie supply (energy intake) as a percentage of requirements, 1989 (*Source:* UNICEF)

---

4 Describe the relationship between the consumption of cassava flour and wheat, compared to income (Fig. 10.9).

10.3 The income factor and nutrition

Income, or purchasing power, has an important influence on people's diet. People are often constrained not only in their choice of food, but also in the quantity and quality they can afford (Fig. 10.8). Certain foods may only be consumed because low incomes prevent a more varied and more appetising diet (Fig. 10.9).

Figure 10.8 Mealtime for a low income British family

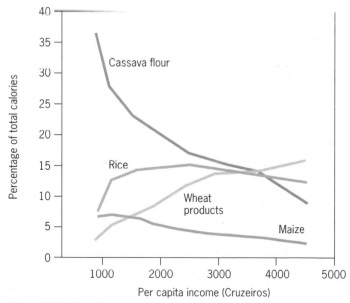

Figure 10.9 Contributors to per capita energy consumption (by income class): North-east Brazil, 1975-6 (*After:* Tabor, 1979)

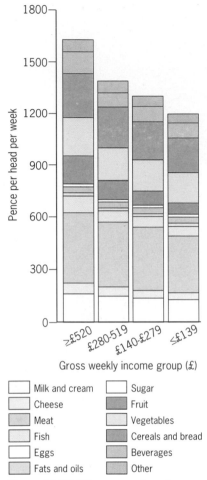

Figure 10.10 UK consumption of various foods by income groups, 1993 (*Source:* MAFF)

Engel's law states that as incomes increase, the proportion of income spent on food decreases. It is by no means unusual for poor people living in rural areas in economically developing countries to spend 80 per cent of their income on food. This contrasts with a wealthy country like the UK, where on average food accounts for just 15 per cent of total expenditure by each household. There will, though, be variations to such a rule within each country.

However, Figure 10.10 shows that spending on *certain foods* increases with income, while on others spending declines.

Income and diet

If low incomes produce an inadequate diet at a global level, we would expect to see a correlation between GNP and diseases resulting from malnutrition.

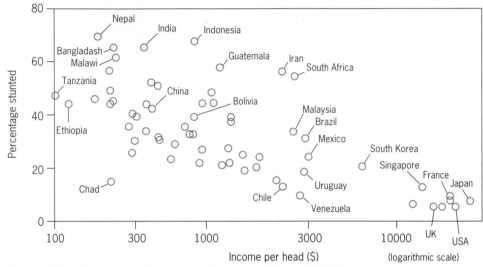

Figure 10.11 Stunting of children's growth by malnutrition, 1980-90

10.4 Nutrition and food-related diseases

Illness due to poor diet occurs in both economically developing and economically developed countries. However, the causes and effects are usually different.

Mal/under/overnutrition

Let's begin with some definitions. Undernutrition is caused by too little food and ultimately will lead to death by starvation. The Food and Agriculture Organisation (FAO) of the United Nations has identified undernutrition among people who do not consume enough food to maintain their body weight and support light activity. Malnutrition results from an unbalanced diet, and a lack of particular nutrients (Fig. 10.12). It is common in the developing world, but it is possible (though less common) to be malnourished in an economically developed country. Death in childhood due to poor nutrition is not common in most economically developed countries. Overnutrition (i.e. overweight and obesity) is undoubtedly more of a problem in richer countries and occurs when a person's diet contains more energy than they use in activity (Fig. 10.13). Some of the diet-related disorders common in developed countries are shown in Table 10.2, and dietary factors have been shown to increase the risk of these disorders. For example, if a person consistently consumes more energy than they need, they will put on weight. Similarly, a diet high in fat increases the risk of heart disease. Obesity is a risk to health, as is being underweight.

?

5a With reference to Figure 10.11, describe and explain the association between the stunting of children's growth through deficient diet and income per head.

b Which countries deviate most from the general trend in Figure 10.11? Suggest reasons for their deviation.

6 Compare and comment on the global distributions of vitamin A deficiency and heart disease (Figs 10.12 and 10.13).

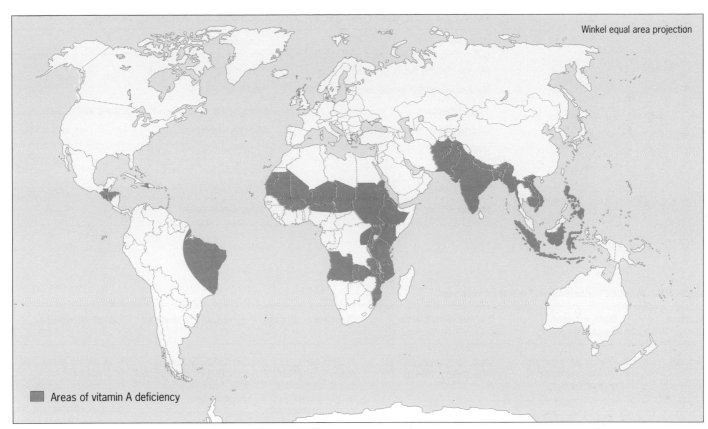

Winkel equal area projection

■ Areas of vitamin A deficiency

Figure 10.12 Malnutrition: vitamin A deficiency (*Source:* FAO)

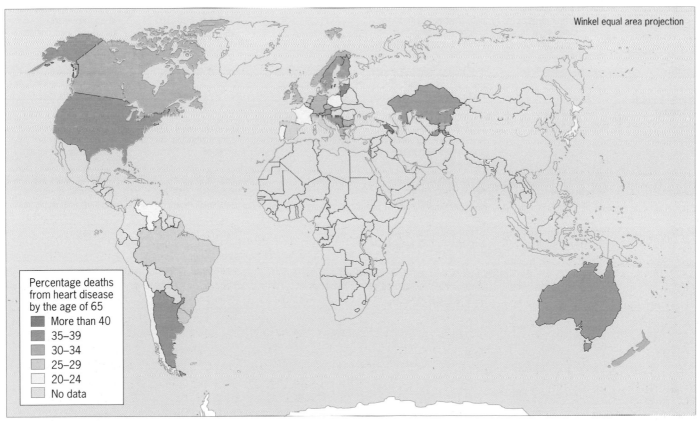

Winkel equal area projection

Percentage deaths
from heart disease
by the age of 65
■ More than 40
■ 35–39
■ 30–34
■ 25–29
□ 20–24
□ No data

Figure 10.13 Deaths from heart disease (partly due to excess fat consumption) (*Source:* WHO)

Table 10.2 Food related diseases common in economically developed countries

Disease	Dietary links
Atherosclerosis (heart disease)	Foods high in cholesterol
Hypertension (high blood pressure)	High salt intake
Type 2 diabetes	Foods high in refined carbohydrates, low in fibre
Diverticular disease (colon)	Insufficient fibre
Hiatus hernia	Insufficient fibre
Gallstones	Foods high in fat, refined carbohydrates
Kidney stones	Foods high in animal protein, refined sugar, low in fibre
Bladder stones	Diet high in cereals

Table 10.3 Nutrient deficiency and resultant diseases

Nutrient	Main nutrient functions	Source of nutrients	Deficiency disease	Symptoms of disease	Extent of problem
Protein	Release of some energy Growth and repair of body tissues	Meat, cheese, eggs, fish, nuts, pulses, cereals	Protein energy malnutrition (PEM), 2 forms: marasmus and kwashiorkor, ultimately starvation	Swollen belly, apathy, muscle wasting, weight loss, flaky skin, hair discoloration	PEM affects about 25% of populations in economically developing countries. About 10m die each year as a result of reduced immunity to disease.
Fat soluble vitamins					
Vitamin A	Essential for vision in dim light: needed for growth and for maintenance of skin and membranes	Milk, cheese, liver and fatty fish. Carotenes in carrots, potatoes, yellow fruits, green leafy vegetables	Xerophthalmia	Poor sight or even blindness, reduced resistance to infection	Affects 50% of children in economically developing countries.
Vitamin D	Calcium absorption	(Formed in skin by) sunlight, dairy produce, oily fish, fortified foods	Rickets in children, osteomalacia in adults	Bone deformities	Insufficient exposure to sunlight e.g. women in purdah, poor diet.
Water soluble vitamins					
Niacin (vitamin B)	Release of energy	Liver and other meats, some grains, groundnuts, dried peas and beans, milk, fortified foods	Pellagra	Loss of weight, skin rash, diarrhoea, mental disorder	Especially maize diets because niacin is unavailable in maize.
Thiamine (vitamin B1)	Release of energy from carbohydrates, nerve functions	Dried peas and beans, whole grains, milk, eggs, vegetables and fruit	Beriberi	Loss of appetite, weakness, swelling (oedema), heart failure	When nutrients are perpetually destroyed by cooking e.g. especially in polished rice areas.
Vitamin C	Production of collagen, wound-healing, iron absorption	Citrus and other fruits, potatoes, green vegetables	Scurvy	Bleeding gums, slow healing of wounds, painful joints, bone weakening	Unknown
Minerals					
Iron	Formation of red blood cells	Liver, meat, vegetables with dark green leaves	Anaemia	Blood disorders causing tiredness, loss of appetite, apathy	Affects 917m, especially pregnant women in economically developing countries. In addition to being caused by low intakes, may be due to an inability to absorb sufficient iron or to particularly high requirements in some groups e.g. young women.

Figure 10.14 Our bodies' basic needs

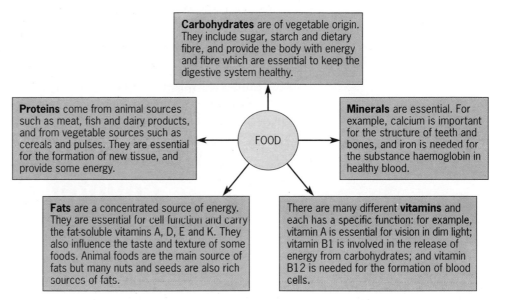

Diet and health are closely linked. We all need a balanced diet of carbohydrates, proteins, fats, vitamins and minerals (Fig. 10.14). An inadequate or unbalanced diet can increase risk from a number of diseases, and in severe cases cause diseases of deficiency or excess (Tables 10.2 and 10.3).

In recent years, many people have become more interested in the idea of a 'healthy' diet. Among the general public in Europe and North America, there are worries about the need to lose weight and to avoid food additives which may cause diseases such as cancer. The British government offers advice to its citizens through its Committee on Medical Aspects of Food Policy (COMA) (Table 10.4). COMA makes recommendations about the intake of energy and nutrients for different groups of the population. The recommendations are based on current knowledge about the links between diet and disease. COMA addresses areas of particular concern to the public:

• Sugar: This is a type of carbohydrate. It provides energy but no other nutrients. Sugar can cause tooth decay if eaten frequently particularly unbound sugar which is found in foods such as honey, table sugar and confectionery. Many sugary foods are also high in energy and fat.

• Fat: A diet containing a lot of fat, particularly saturated fats, is known to be associated with a raised blood cholesterol level. This is one of the risk factors for heart disease. Fat is a concentrated source of energy, so a high fat diet often provides a lot of energy and may cause people to put on weight.

Table 10.4 COMA's proposals for a healthy diet

Nutrient	Examples of sources	Average intake, 1992	COMA recommendation
Total fat	Butter, margarine, oils, meat, biscuits, cheese	42% total calories	Not more than 35% total calories
Saturated fat	Meat, palm oil, lard, butter, cheese	16% total calories	Not more than 10% total calories
Carbohydrates (complex)	Bread, pasta, potatoes, rice, lentils and dried beans, fruits	31% total calories	39% total calories
Sugar (unbound)	Sweets, chocolate, biscuits, cakes, jams	14% total calories	Not more than 11% total calories
Fibre	Bran, wholemeal bread, brown rice, fruits, vegetables	12g a day	Range 12–24g/day
Protein	Meat, cheese, eggs, nuts, pulses	13.5% total calories	10–15% total calories

The main sources of saturated fats are found in meat and meat products, milk and milk products such as butter and cheese, and hard margarine.

- Salt: A high salt diet may increase the risk of high blood pressure.

- Fibre: A diet low in fibre may increase the risk of some diseases. Fibre is essential for the efficient functioning of the digestive system.

?

7 Read the articles in Figures 10.15 and 10.16.

a Draw a table to identify:
- how the average British diet may be thought of as unhealthy,
- why so many people in the UK could be considered to have unhealthy diets.

b Using your table, write a brief introduction to Suzi Leather's report (Fig. 10.16) explaining how an unhealthy diet may be linked to disease. Give your own suggestions as to how people's diets could change if they are to reduce the risk of disease.

8 There are numerous food consumption patterns across the world (Fig. 10.1). Using your survey from question **1**, consider people's attitudes and concerns over food and write an article explaining different preferences.

9 Glasgow is described as 'the coronary capital of the world' (Fig. 10.15). Many people in the developed world can decide how and what they will eat, while many in economically developing countries have little choice.

a Keep a diary for most of your meals over the following week. List the main foods and nutrients each meal contains.

b Analyse your eating habits according to the COMA recommendations (Table 10.4).

c Compare your meals with others in your group.

d Discuss if and whether, when we have enough information to make effective decisions, we continue to make poor choices over what we eat.

'Peasant diet' urged to counter disease

A RETURN to a 'peasant diet' was urged yesterday after poor eating habits were blamed for making half the middle-aged population overweight and 10 per cent of adults medically obese.

At the launch of a World Health Organisation report calling for a drastic reduction in the amount of sugar and animal fats we consume, Professor Philip James said: 'Britain has a very, very poor health record when it comes to diet related diseases'.

The report, which sets world-wide nutrition guidelines, is a radical attack on the main-stream food industry because it recommends a return to a 'peasant diet' based on simple, less refined foods with strong emphasis on fruits, vegetables and complex carbohydrates — bread, potatoes, rice, pasta and pulses. Professor James, who chaired the team producing the WHO report, described Glasgow as the 'coronary capital of the world' and blamed the UKs high rates of breast and colon cancer on poor eating habits.

The sugar industry takes the worst hammering with a recommendation that refined or free sugars (both brown and white) should make up no more than 10 per cent of calories consumed daily and could safely fall to 'zero' because they 'contribute no nutrients and are not essential for human health'.

Sugar makes up at least 18 per cent and probably 25 per cent of calories in the typical British diet, according to the latest Ministry of Agriculture survey.

Fat currently makes up 41 per cent of calories in the British diet. The WHO report says fat should fall to 30 per cent with strict limits on the cholesterol derived from meat and milk products. The lower limit for cholesterol is set at zero, which means that vegan diets without meat, eggs and dairy products can be healthy.

Figure 10.15 Nutrition guidelines for a healthy heart (*Source:* James Erlichman, *The Guardian*, 19 April 1991)

Healthy food 'too expensive for poor'

P OVERTY, not ignorance, forces many people to eat unhealthy food, according to a report published yesterday. The report was compiled after its author, Suzi Leather, challenged the Ministry of Agriculture to produce a healthy weekly diet which poor people could afford.

The Government had refused all previous demands that it costs more money to provide a decent diet, despite allegations that low state benefits force many people to feed their children biscuits and crisps because they cannot afford healthy food.

The ministry's 'low-cost healthy' diet costs £10 a week — roughly a third of income support for many people. But it allows only one fresh egg every two weeks, excludes yoghurt, and relies heavily on tinned vegetables, bread and breakfast cereals.

'Of the eight slices of bread to be eaten each day, only three would have even a thin spread of margarine or butter — the rest would have to be eaten dry', Ms Leather said.

The real picture was worse because inadequate benefits forced many families into debt and they had to cut back further on food.

The ministry's diet also assumes that families have cooking facilities, although many, especially in bed and breakfast accommodation, do not.

Figure 10.16 Poverty forces unhealthy eating (*Source:* James Erlichman, *The Guardian*, 6 Nov. 1992)

Measuring poor nutrition

It is not easy to measure undernutrition and malnutrition. Childhood malnutrition causes growth failure which means the weight of a malnourished child is less than the ideal for its age. Weight and age are therefore frequently used to assess malnutriton by comparison with a standard.

The choice of standard can be controversial: for instance, the 'small but healthy' controversy arose when Western nutritionists compared children's growth in economically developing countries with the USA (Fig. 10.17). The analysis measured undernutrition by comparing a child's actual weight for height with the US average. However, some nutritionists believe that Western standards are not appropriate for all populations. The diets of poor people in the Tropics contain less protein than in the USA, and consequently children grow more slowly. We therefore have no basis for assuming that protein-rich western diets are superior, and thus no reason for describing a small but healthy child as 'malnourished'.

Figure 10.17 Median male weight-for-age in seven Latin American countries and the USA (*After:* US White House, President's Science Advisory Committee, 1967)

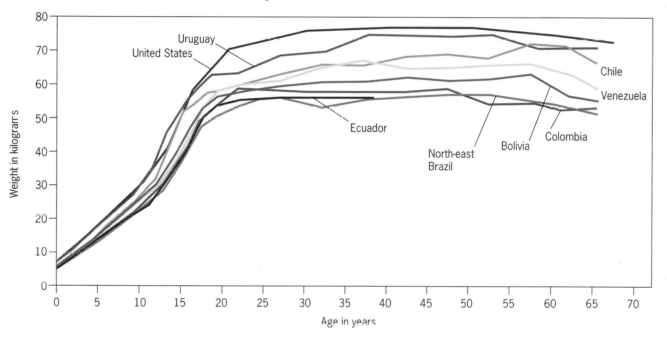

In the 1960s, protein deficiencies were seen as a major cause of malnutrition in the economically developing world. But by the mid–1970s this view had changed. Now we believe that if a person consumes enough energy (calories), then they will probably consume sufficient protein. The exception to this are diets dominated by starchy staples such as those based on cassava, yam or plantain in West Africa. All of these foods are low in vegetable protein.

10.5 Famines

In recent years we have all been shocked by pictures of severely undernourished children in Africa on our television screens. Yet famine is nothing new, and at one time or another has probably affected most parts of the world (Table 10.5). In England famines occurred regularly before 1600; in Scotland they continued until the end of the seventeenth century; and in Ireland there were terrible famines in 1740–1, and in 1845–50. The latter was known as the 'Great Hunger', and is the subject of the case study on pages 139–140.

Table 10.5 Population losses from major famines

Location	Dates	Estimated deaths (millions)
India	1837	0.8
Ireland	1845	0.75
India	1863	1.0
India	1876–8	5.0
East China	1877–9	9.0
China	1902	1.0
China	1928–9	3.0
USSR	1932–4	4.0

We all think we know what famine is. A common definition is that 'famine is a widespread food shortage that leads to a sharp rise in regional mortality'. However, we need to refine such a definition because:

1   Famines are not always widespread. Geographically they can be very localised and may affect only one social group or class.

2   Famines are not always the result of food shortage. They can occur when there is a breakdown in the marketing system or simply when people cannot afford to buy the food which is available (Fig. 10.18).

3   Starvation is rarely the actual cause of death. Undernutrition reduces the body's resistance to infection and people may die from a wide range of diseases.

We can identify several types of famine according to their causes and their impact on a population. Sometimes famines occur when population grows faster than food production. This is especially likely in areas of rapid population growth and low agricultural potential. Natural disasters and wars are often important contributory causes of famine (Fig. 10.19). Famines may also result from a breakdown in transport systems, especially if an area relies on food imports. This happened in the Netherlands at the end of World War 2. Finally there are famines which strike only certain sections of the population. Especially vulnerable are the young, the old, and the poor. The Irish potato famine illustrates this selectivity: while the peasants starved, the larger farmers and landowners were unaffected.

Figure 10.18 Distributing food during the 1925 famine in Soviet Russia – the famine was forced by Stalin for political gain

?

10a   Study Figure 10.19 and explain the causes of famine in the Horn of Africa.
b   In spring 1994 what signs were there of impending famine in the Horn?
c   How have people responded to the threat of famine in the Horn?

11a   Research newspaper articles for: • the causes of other recent famines in Africa, • the source and success of aid in the areas of deficit.
b   Assess whether these areas show a changing pattern in the characteristics of famine.

It is starting again. Ten years after the great Ethiopian famine shocked the world, the first grim harbingers of what could become a still worse disaster, if the world closes its eyes, are stalking the countryside. Starvation deaths are being recorded. Figures of severe malnutrition among children have shot up. Precious oxen are being sold for bags of grain. Families are starting to migrate.
Although a few teasing spring showers have brightened the Abyssinian hills with a deceptive wash of green, the sad reality is that, yet again, the rains are far too little and too late. Another hungry season has begun.

The Economist, 30 April 1994

According to the US Agency for International Development, in April 1994 more than 17m people in Ethiopia, Eritrea, Sudan, Somalia and Kenya are at risk of starvation.

Famine reached a peak in the horn in 1985, when thousands died of starvation. Rapid population growth could make the situation far worse in future. The combined populations of Ethiopia and Eritrea in 1985 were 42m; today they are 55m.

Sudan The civil war is the main cause of food shortages. 5m are at risk. In the south, 3.5m have fled their homes and warring factions will not allow food to pass through the war zone. Elsewhere, drought has sent the price of sorghum soaring. A centre for famine relief has been set up at El Obeid.

Eritrea Three decades of civil war, coupled with drought, caused an endless cycle of food shortages. Although the government has taken steps to alleviate food shortages (distributed seeds, oxen, ploughs; planted trees; built dams and terraces and expanded the cultivated area by 25%) unreliable rainfall, locusts and rats have undermined their efforts. Crop failure for 1993 was estimated at 80%. Thus two-thirds of the population depend on food aid.

Ethiopia In Ethiopia, 7m people will need food aid to survive in 1994. Feeding centres have been set up in Tigray and Omo where kwashiorkor and marasmus (PEM) are widespread among children. Farmers are feeding cactus to their cattle, desperate families are eating the seeds they should be planting.

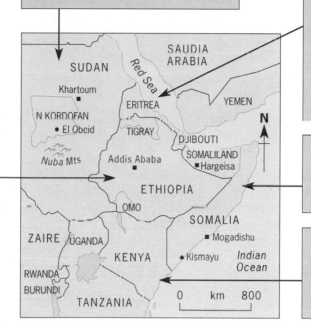

Somalia 750000 depend on food aid. There is plenty of food, but civil war and banditry make distribution of food aid difficult.

Kenya 3m people are at risk. Pastoralists have lost hundreds and thousands of their livestock, as water sources and pasture have dried to dust.

Figure 10.19 Famines in the Horn of Africa, 1994 (After: The Economist, 30 April 1994)

The Irish potato famine

The 'Great Hunger' of 1845–50 in Ireland (Fig. 10.20) killed one million people, or nearly one in eight of the population. It was caused by the failure of the potato crop, which by the early nineteenth century had become the staple food of half the Irish population. The risk associated with dependence on a single crop was highlighted in 1845, when a fungal disease – potato blight – spread rapidly. The resultant poor harvests of 1845, 1846 and 1848 led to widespread starvation and great suffering (Fig. 10.21).

While the 1845 potato harvest was poor, the 1846 crop failed disastrously. Average yields fell from 2.5 to 0.5 tonnes per hectare, and potato prices quadrupled in the space of a few months. Meanwhile the wages of farm labourers did not keep up with prices. They were

therefore unable to feed themselves and their families. The British government responded by introducing a public works programme to provide jobs. Although the scheme employed 750000 people, it was unsuitable for the needs of the Irish population. Instead of focusing on projects such as land reclamation and land drainage to increase food output, the scheme mainly concentrated on road building. In 1847, soup kitchens provided temporary relief but the food crisis continued for three more winters, with catastrophic effects. Mortality was highest among children and old people, and the remote West and South-West Ireland were hardest hit (Fig. 10.22).

The Irish potato famine might seem like a classic

Irish potato famine

Figure 10.20 Ireland

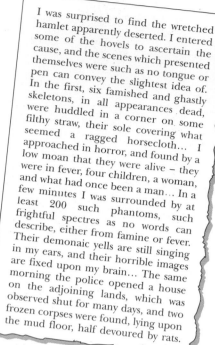

Figure 10.21 Sketch in a house in Fahey's Quay, Ennis, 1850

I was surprised to find the wretched hamlet apparently deserted. I entered some of the hovels to ascertain the cause, and the scenes which presented themselves were such as no tongue or pen can convey the slightest idea of. In the first, six famished and ghastly skeletons, in all appearances dead, were huddled in a corner on some filthy straw, their sole covering what seemed a ragged horsecloth... I approached in horror, and found by a low moan that they were alive – they were in fever, four children, a woman, and what had once been a man... In a few minutes I was surrounded by at least 200 such phantoms, such frightful spectres as no words can describe, either from famine or fever. Their demoniac yells are still singing in my ears, and their horrible images are fixed upon my brain... The same morning the police opened a house on the adjoining lands, which was observed shut for many days, and two frozen corpses were found, lying upon the mud floor, half devoured by rats.

Figure 10.22 Nicholas Cummins' visit to Skibbereen, County Cork (*Source: The Times*, Dec. 1846)

case of population growth outstripping the increase in food supplies. After all, Ireland's population had trebled in the previous hundred years, and its population density per hectare of cultivated land was one of the highest in Europe. Yet this interpretation is too simple and overlooks some important facts. For instance, throughout the years of crisis, food exports to England continued. Also, the government's indifference and incompetence, believing that market forces would eventually provide a solution, made matters worse. Indeed one Irish nationalist was moved to observe that 'the Almighty sent the potato blight, but the English created the famine'.

The massive decline in Ireland's population after 1845 was not only due to mortality caused by famine.

Many people decided to migrate, principally to the UK and the USA (Fig. 10.23). In some parts of rural Ireland, depopulation was so severe that even today the population has not recovered to its pre–1845 levels: in 1831 the population was 7.8 million, while in 1901 it was only 4.5 million. In contrast, the population of England and Wales jumped from 13.9 in 1831 to 35.5 in 1901.

?

12 Use Figures 10.21, 10.22 and 10.23 to write an account of life in Ireland in the late 1840s. Imagine you are a poor farmer, or an emigrant to America, and describe your experiences.

Figure 10.23 Irish emigrants boarding ships for Liverpool, 1851

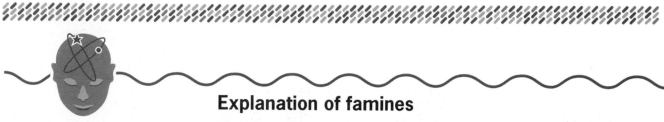

Explanation of famines

The economist Amartya Sen has put forward a theory which suggests that famines are not always caused by a decline in **food availability (FAD)** – a decrease in the food supply. He argues that they are the result of a deterioration in the **entitlements** of certain sectors of society. In simple terms this means that poor people have limited access to food because of their weak purchasing and bargaining power.

A person's entitlements are made up several things. They might include the ownership of land, or occupation and status. For example, a civil servant in an economically developing country might, because of his/her occupation, be given a food ration priority. But for the majority of people, it is exchange entitlements which are important. These are usually a matter of income, and earning sufficient money to buy food to prevent undernutrition.

Those people who depend wholly on exchange entitlements are particularly at risk in times of food shortage when food prices rise rapidly (Fig. 10.24). The impact of famines, whatever their cause, is always to widen the gulf between the 'haves' and 'have nots'. The rich buy up the assets of the destitute who are forced to sell at knock-down prices to buy food.

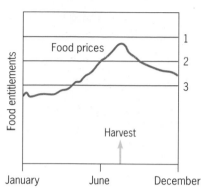

Figure 10.24 Food prices and food entitlements

?

13 Study Figure 10.24 and Table 10.6.

a Suggest two factors which might account for the inflation in food prices.

b Why do food prices peak in July?

c Describe and explain the variable impact of food prices on the three households in Table 10.6.

Table 10.6 Household characteristics and entitlements

Status/income	Land	Entitlements/other assets	Savings/stored food
1 Farmer: above average	Owner-occupier of 10ha	Livestock, oxen, ploughs and other means of production	Significant savings
2 Smallholder: average income	Tenant; fixed rent; farms 2ha	Owns a few livestock (pig, poultry). All means of production owned by landlord	Small savings
3 Landless labourer: low income	None	None	None

Famine in Bangladesh, 1974

Background

Bangladesh (Fig. 10.25) has a history of famine. The 1940–3 famine in Bengal (now mainly Bangladesh) was indirectly responsible for ten million deaths. This tragedy was repeated, though on a smaller scale, in 1974. The economic status of the rural poor in this part of South Asia has gradually worsened over the last 50 years. The value of wages has declined steadily, and the average size of land holding has fallen. The latter reflects inheritance practices (the sub-division of farms between heirs on the death of the landowner) and the dispossession of smallholders who have fallen into debt. Thus the number of sharecroppers and landless labourers has increased, with a corresponding decrease in owner-occupiers.

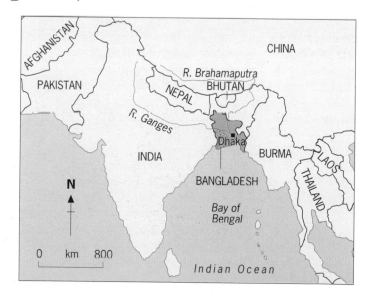

Figure 10.25 Bangledesh

Famine in Bangladesh

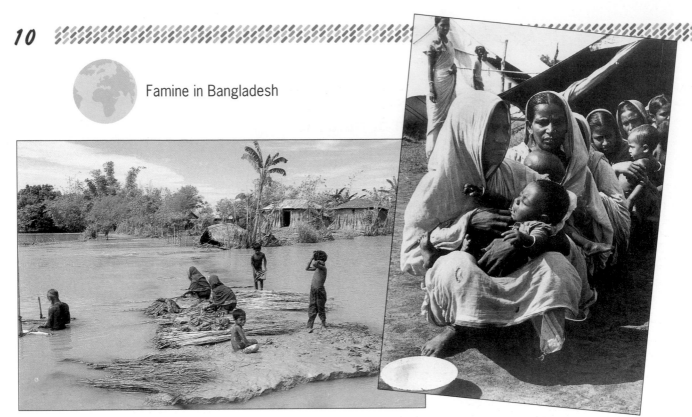

Figure 10.26 Jute stripping: work continues despite monsoon floods over Bangladesh farmlands

Figure 10.27 Queuing for food aid during the famine in Bangladesh, 1974

The natural environment of much of Bangladesh is hazardous to farming. Most of the country comprises the delta of the Ganges, Brahmaputra and Meghna rivers. Low-lying and close to sea-level, Bangladesh is vulnerable to flooding not only by rivers but also by tidal surges driven by tropical cyclones in the Bay of Bengal (Fig. 10.26). Every year, during the monsoon season between May and September, half the country floods to a depth of at least 30 cm, and one third to a depth of two metres. July 1974 brought the worst floods in living memory, causing further dislocation to a transport network and food supplies already wrecked in the war of independence from Pakistan three years earlier (Fig. 10.27).

Thus in 1974, before the onset of famine, environmental and political catastrophies had already left Bangladesh in a very weak condition. Food availability decline is therefore a tempting explanation of the 1974 famine, especially since food consumption among the poorest 12 per cent of Bangladeshis averaged just 863 kcals a day.

The progress of famine

The first signs of famine occurred in the period between February and June, 1974. People began eating alternative 'famine foods' like plantain saplings and banana leaves. These foods even started to appear on the market. There was also a noticeable increase in thefts and begging, and large numbers of people flocked into the towns, especially the capital Dhaka, looking for relief.

The famine peaked in the months July to October – a similar timing to other devastating Bengal famines, such as those of 1770, 1866 and 1943. Up to two million people migrated from their homes to find food, some walking as much as 160 kilometres. Social and family bonds often disintegrated: parents abandoned their children, and husbands deserted their wives. Desperate people sold or mortgaged their land; their cattle and agricultural implements, and even their household utensils. In fact, anything that might help them to buy food in the short term was sold.

By November, the worst was over but the Bangladeshi government had been too slow to cope with the crisis. They had not declared an emergency until September 1974, when they set up 4300 soup kitchens (langarhhanas), but these had been closed by late November.

Table 10.7 Mortality (per 1000) in famine areas by occupational groups: August to October 1974 (Source: Alamgir M, 1980)

Occupation	Adults	Children
Farmers (size of property)		
<0.2 ha	100	148
0.2–0.4	79	142
0.4–1.0	46	78
1.0–2.0	9	12
Wage labour	129	201
Trade	27	38
Services	26	28
Other	35	0

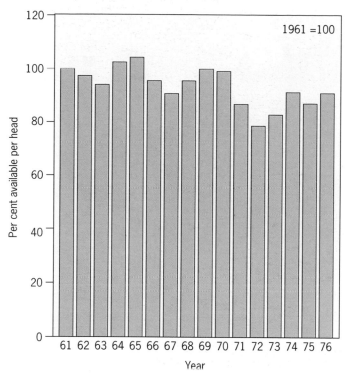

Figure 10.28 Bangladesh: rice production, 1961-76

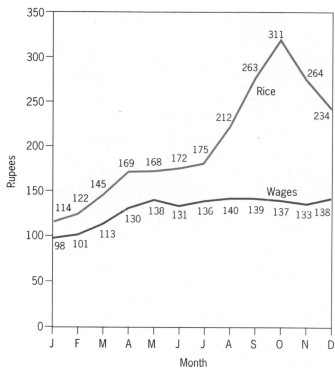

Figure 10.29 Bangladesh: relationship between wage and rice retail price indexes, January-December 1974 (*Source:* Amartya Sen, 1981)

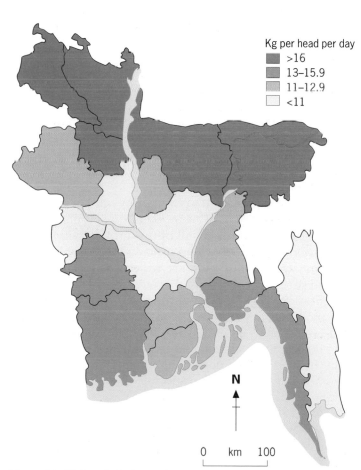

Kg per head per day
- >16
- 13–15.9
- 11–12.9
- <11

N

0   km   100

Figure 10.30 Bangladesh: availability of food, July-October 1974

?

14 Using the ideas of Amartya Sen and food entitlements, suggest an explanation for the mortality pattern in Table 10.7. Test the hypothesis that the Bangladeshi famine resulted from FAD, or shortage of food within the country, by completing the following tasks:

15 Study Figure 10.28 and comment on the size of the 1974 harvest.

16a Using the data in Figure 10.29, divide the rice price index by the wage rate index. (The resulting exchange rate is a measure of the purchasing power of consumers.)
b Describe what happened to the retail price of rice and the wages of labourers and small farmers during 1974.
c What effect would these trends have?

17 Study Figures 10.30–10.32. Figure 10.30 shows us the areas where food was in shortest supply during the famine. Figure 10.31 is a measure of the spatial pattern of famine (i.e. where food needs were greatest). Figure 10.32 tells us about food entitlements, and specifically where purchasing power had declined most.
a Compare Figures 10.30 and 10.31. Do the areas of famine correspond to the areas of greatest food shortage?
b Compare Figures 10.31 and 10.32. Do the areas of famine correspond to the areas of greatest entitlement decline?
c Using all of the evidence in Figures 10.28–10.32, comment on the validity of the hypothesis that the Bangladesh famine resulted from FAD.

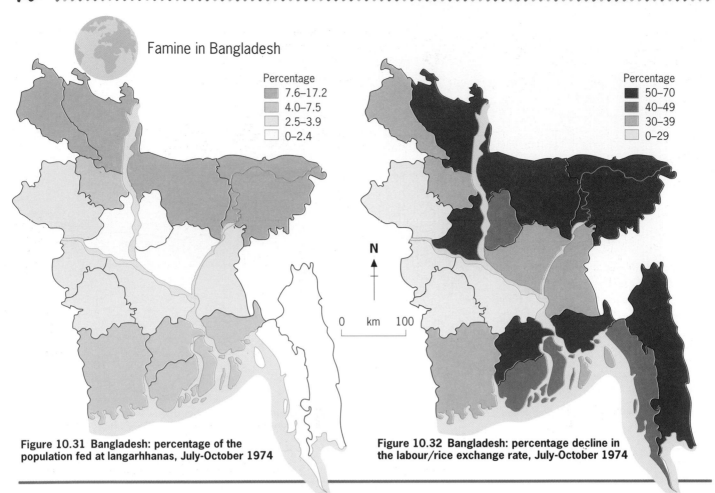

Famine in Bangladesh

Percentage
- 7.6–17.2
- 4.0–7.5
- 2.5–3.9
- 0–2.4

Percentage
- 50–70
- 40–49
- 30–39
- 0–29

N

0   km   100

Figure 10.31 Bangladesh: percentage of the population fed at langarhhanas, July-October 1974

Figure 10.32 Bangladesh: percentage decline in the labour/rice exchange rate, July-October 1974

10.6 Future trends

Overall, average levels of nutrition and food production (Fig. 10.33) do seem to be rising in the right direction. Table 10.8 shows that 89 economically developing countries had average food consumption of less than 2500 kcals per head in the early 1960s. This should be reduced to 46 by the year 2000 – still an unacceptably high number, but a vast improvement.

Rising world demand for food

Latest forecasts suggest that by 2020 the world's population will have reached nearly eight billion. Only by the year 2050 will the population stabilise, somewhere between 10 and 12 billion. Virtually all of this growth will take place in the economically developing world, where already food supplies are inadequate and undernutrition and malnutrition are widespread. Hardly surprising therefore, that the success of rising food production of the last 30 years is little cause for celebration. What really counts is not increases in total food production, but increases in per capita food production.

World demand for food is also set to increase for reasons other than population growth. Rising incomes will almost certainly mean changes not only in the amount, but in the type of food consumed. A shift from staples such as roots and cereals to higher value foods (especially livestock products) is likely. As we saw in Chapter 3, livestock farming is a very wasteful means of producing food. In the early 1980s, 72 per cent of grain consumed in economically developed countries was fed to animals, compared to just 13 per cent in the economically developing world. Any further increase in meat consumption among the wealthy in developing countries would reduce the amount of grain left for poor, rural dwellers.

Table 10.8 Number of economically developing countries at different average levels of nutrition (kcals per head) (*Source*: Alexandratos and FAO)

	1961/3	1969/71	1979/81	1988/90	2000 projected
<1900	23	7	8	8	4
1900–2500	66	74	54	56	42
>2500	5	13	32	30	48

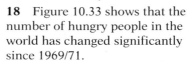

18 Figure 10.33 shows that the number of hungry people in the world has changed significantly since 1969/71.

a Calculate:
• the absolute change,
• the percentage change, in the number of chronically undernourished people between 1969/71 and 2000.

b What proportion of the world's population was undernourished in 1969/71 compared to 1988/90?

c Which continent has experienced:
• the largest increase,
• the largest decrease, in the chronically undernourished?

19 Essay: If you were an economist working for the FAO, would you be optimistic or pessimistic about the prospects of eliminating hunger in the developing world over the next 20 years? Justify your answer.

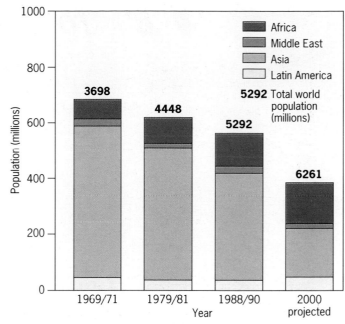

Figure 10.33 Number of chronically undernourished people in economically developing countries.

Summary

- A healthy diet requires a balanced intake of sufficient protein, carbohydrate, fats and lipids, vitamins and trace elements.

- Food consumption and diet are influenced by age, gender, occupation and metabolic rates, as well as income and culture.

- There are spatial patterns of diet and food preferences at local, regional, national and global scales.

- As incomes rise, the proportion spent on food decreases, though spending on certain foods (e.g. animal products) often increases.

- People's health is closely tied to diet and food consumption. Too little food causes undernutrition; an unbalanced diet leads to malnutrition. Both are common in economically developing countries and leave millions of people susceptible to disease.

- In economically developed countries, overnutrition is widespread and is a cause of several chronic, degenerative diseases, such as heart and circulatory problems.

- Famine and hunger affect up to one in seven of the world's population.

- Today famine and hunger are concentrated in the economically developing world but in the past they have affected nearly all parts of the world.

- Famines are socially, economically and geographically selective in their impact.

- There is no global shortage of food, only local and regional shortages caused by production and distribution problems. These problems are frequently linked to local or regional political, economic and social conditions.

- The Bangladesh famine in 1974 was not caused by food availability decline, but by a loss of food purchasing power (entitlement) by the poor.

- There have been significant increases in world food production in the last 30 years, which have outstripped the increase in world population.

- One of the greatest threats to food supplies and adequate nutrition is the world's rapidly expanding population.

11 Some solutions to food problems

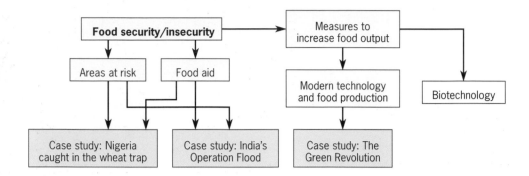

11.1 Introduction

One thing must be made clear at the outset: there is no global shortage of food at the moment and there is not likely to be one in the near future. With current levels of food production, we could feed all of the world's population with an adequate and balanced diet. Where food shortages have arisen in recent years, they have often been caused by poverty and local problems of food production and distribution. There is no reason to believe, as some writers suggest, that the planet's agricultural resources have reached their limit. Instead, food shortages are best understood as a failure of economic and political systems to make food available where and when it is needed. We shall first of all examine the concept of **food security**, then look at two approaches to solving famine and hunger – through food aid, and through modern technology to boost food production.

11.2 Food security

The concept of food security has become influential since the early 1970s. People are secure in their food supply only when they have access to sufficient food to lead a healthy life. Essentially, this means that food must be available in adequate amounts, and that its supply must be assured. However, food security is more than the simple *availability* of food. Food can only be bought if it is sold at prices people can afford. Moreover, food may be available within a country or region but without adequate transport, storage facilities and markets people may have no access to it. We saw in the previous chapter that although there is no overall shortage of food in the world, famine and hunger are all too common. The implication is that hundreds of millions of people experience chronic food insecurity.

Improving food security

Improvements to food security have been made possible since the 1960s (Fig. 11.1). Several factors account for the growth in global food supply:

1   A massive increase in the use of fertilisers in low income countries.

2   New agricultural technologies, especially improvements in plant yields associated with the **Green Revolution** (see Section 11.4).

3   An expansion of the cultivated area and multiple cropping of existing land.

4   An expansion of the area of irrigated cropland.

Figure 11.1 Projected increases in crop yields

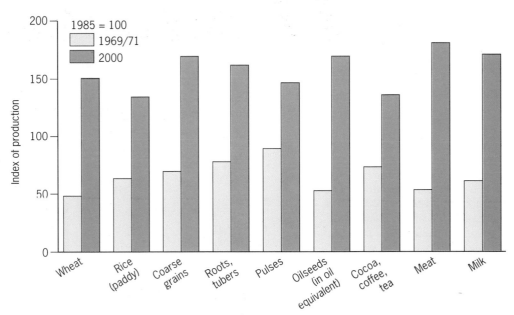

?
---

1 With reference to **intensification** and **extensification**, explain how the activities in Figures 11.2 and 11.3 could increase food production.

Table 11.1 The world's resources of cultivable land (*Source*: Buringh and Dudal, 1987)

	Million hectares	Percentage
Total land area	13 392	100
Cultivated land	1461	11
Potential for cultivation	1570	12
Not cultivable	10 361	77

Increasing the area of cultivated land

Approximately one quarter of the world's land surface could be used for cultivation. As it is, only 11 per cent is actually cultivated (Table 11.1). Thus three-quarters is either of low fertility and suitable only for rough grazing or forestry, or has some physical constraint such as aridity, extreme cold, and adverse relief. The greatest potential for expanding the cultivated area is in Africa, which currently uses only 21 per cent of its potential arable land. Comparable figures for South America and Central America are 15 per cent and 49 per cent respectively.

Figure 11.2 Extensification: Australian sheep-grazing land

Figure 11.3 Intensification: irrigated market gardening with olive groves, Crete

2 Governments adopt a range of policies to secure their nation's food supplies (Fig. 11.4). Refer to the index and relevant sections in this book and explain how food security is linked to the factors shown in Figure 11.4.

3 Describe the global pattern of food security (Fig. 11.5) within the developing world. From what you have learnt about agriculture and food in previous chapters, make a list of environmental, economic and social factors that might be associated with the countries which have low food security status.

How many people could the earth support at adequate levels of nutrition? A recent major research project tried to answer this question. It used detailed climatic and soil data, known crop requirements, and made assumptions about technological change, population growth, and carrying capacities. It concluded that if all potential cultivable land were used, a population of around 8 billion could be supported. However, the regions of greatest agricultural potential and highest population density do not necessarily coincide.

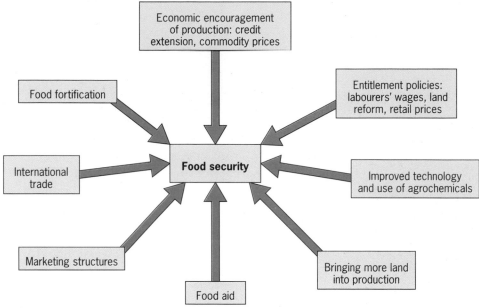

Figure 11.4 Factors in national food supply

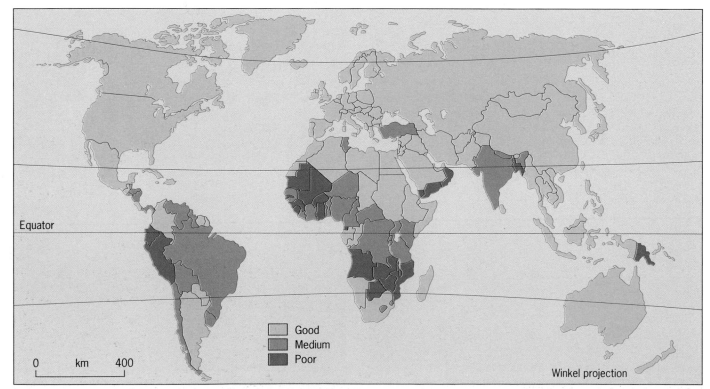

Figure 11.5 Global food security status (*Source:* Jazairy, 1992)

Table 11.2 Regional food production 1961–89 (million million kcals/day) (*Source*: FAO)

	1961	1970	1980	1989
Africa	598	790	1118	1472
North and Central America	810	997	1218	1425
South America	362	484	648	781
Asia	3143	4383	5956	7650
Europe	1378	1566	1727	1801
Oceania	47	57	68	80
USSR/CIS	673	812	898	974
World	6999	9076	11 619	14 160

Tabe 11.3 Regional population change 1961–89 (millions) (*Source*: FAO)

	1961	1970	1980	1989
Africa	287	362	477	623
North and Central America	275	321	374	423
South America	151	191	241	291
Asia	1699	2102	2583	3055
Europe	488	480	505	521
Oceania	16	19	23	26
USSR/CIS	218	243	266	289
World	3094	3718	4469	5225

?

4a Use the data in Table 11.2 and Table 11.3 to calculate average regional food consumption per head for 1961, 1970, 1980 and 1989. You can do this by dividing each figure in Table 11.2 by the equivalent figure in Table 11.3 and multiplying each answer by 1000. This will give you the average number of kilocalories consumed per person.
b Plot the average food consumption trend for each region on a graph like Figure 11.6, and then use it, together with Tables 11.2 and 11.3, to answer the following questions:
c In which three regions in 1989 was average food consumption per head below the world average?
d In which region in 1989 were the people on average worst fed?
e Between 1980 and 1989 Africa's food production increased by nearly one-third, and yet food consumption per head remained more or less static. Why was this?
f Which region made the most rapid progress between 1961 and 1989 in raising average food consumption? By referring to Tables 11.2 and 11.3 suggest two possible reasons for its success.
g Which regions are most likely to give cause for concern if present trends in food consumption continue?

Areas at risk

The study also looked at 117 economically developing countries and concluded that 55 were currently 'critical' because of their inability to feed their populations adequately given their prevailing low technology. Most worrying were the 19 countries identified as 'critical' even if their technology could be brought up to a high level.

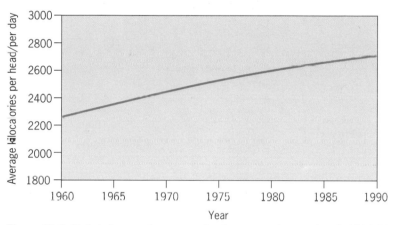

Figure 11.6 Global changes in average food consumption per head, 1961-89

11.3 Food aid

We assume that the international community provides food aid to areas at greatest risk of food insecurity (Fig. 11.7). But have you ever thought about the motives for providing food aid? We might justify food aid on moral

Figure 11.7 Care International supplying corn from the USA as food aid to Angola, December 1991

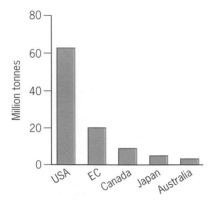

Figure 11.8 Major food aid donors, 1980/81–1989/90

grounds; that rich countries should help those which are less fortunate (Fig. 11.8). Or, more pragmatically, we might argue that, because food is a daily need, food aid might help people over a temporary crisis until they are able to help themselves again.

Sometimes, however, there are 'hidden' reasons for food aid. Some countries might act out of self-interest, giving food aid to create a market for their surplus produce. The government of a donor country thinks that, when the free aid stops, the recipient will want to continue the supply by purchasing food. There are also political considerations: food aid is often used to support friendly governments whether their need is real or not.

Table 11.4 Main cereal food aid recipients, 1987/8–1989/90. (*Source*: FAO and World Bank)

	Food aid (thousand tonnes)	Aid/million population (tonnes)	GNP/head 1989 ($)	Calories/head (1988/89)
Population >5 million				
Bangladesh	3850	36 288	180	1925
Egypt	4282	85 477	640	3213
Ethiopia	1939	43 286	120	1658
Pakistan	1506	14 691	370	2200
Mozambique	1407	93 363	80	1632
Tunisia	1178	154 934	1260	2964
Sudan	1151	49 805	330	1996
India	987	1238	340	2104
Population <5 million				
Cape Verde	160	465 407	780	2778
Jamaica	738	307 417	1260	2558
Guyana	146	183 688	340	2495
Sao Tomé and Principe	20	176 522	340	2153
Costa Rica	379	145 615	1780	2782
El Salvador	623	127 184	1070	2415
Dominica	10	127 160	1800	2911
St Kitts and Nevis	5	113 636	3000	2435

———— **?** ————

5 Using the information in Table 11.4 test the following hypotheses:
Hypothesis 1: most food aid goes to the poorest countries.
Hypothesis 2: most food aid goes to countries with the lowest levels of nutrition.
a Draw scattergraphs for each hypothesis. For hypothesis 1 plot food aid per million (y) against GNP/head (x). For hypothesis 2 plot food aid per million (y) against calories/head (x).
b Calculate the Spearman rank correlation coefficient and its significance level (Appendix A1) for each of the two pairs of data.
c Write a paragraph clearly setting out your main conclusions.

Food aid – who benefits?

There are two aspects of the present system of food aid which attract criticism. First, as economically developing countries import cheap foreign food, they can undermine their own local agriculture which may be unable to compete in the market. Second, food aid can cause the receiving country to become unhealthily dependent on that aid. It is then the donor country which is likely to benefit more than the recipient by continuing the supply of food.

Nigeria: caught in the wheat trap

In many economically developing countries, wheat first became available through food aid. It proved popular, and bread and wheat products made from commercial imports quickly became an essential part of people's diets, usually at the expense of traditional staples. This process has been called the 'wheat trap'. Nigeria (Fig. 11.9) is a good example of this problem. In the early 1980s Nigeria imported 1.5 million tonnes of wheat a year (Fig. 11.10). This was costly in foreign exchange, and so the government tried to replace imported US wheat with home-grown wheat. To grow wheat required investment in **irrigation** schemes as

few areas of Nigeria have a suitable climate for rain-fed wheat cultivation (Fig. 11.11). This attracted wealthy transnational agribusiness corporations.

The cost of home-grown wheat

Nigerian wheat is expensive to produce, and much of the technology has to be imported. This is a further drain on Nigeria's foreign exchange which the country can ill afford. None the less, bread is popular in all sections of society, and especially among the urban poor. As a result a network of flour mills and bakeries was established to meet the growing demand.

Figure 11.9 Nigeria

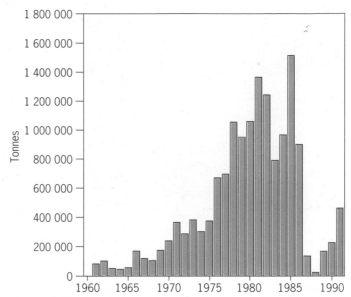

Figure 11.10 Nigeria: import of wheat and wheat flour, 1960-91 (*Source:* FAO)

Figure 11.11 Harvesting wheat on the South Chad Basin Development Authority Irrigation Project, Borno District, Nigeria

Figure 11.12 Nigerians queuing for wheat rations

All of this has been extremely wasteful of resources. Government investment would have been better spent on developing rain-fed agriculture. Not only would this have helped small farmers, it would have encouraged traditional crops, such as maize or sorghum, which are better suited than wheat to the Nigerian agro-climate and diet.

Expanding production

The oil boom, experienced in the Nigerian economy during the 1970s, provided the government with money to invest in wheat cultivation. However, falling oil prices in the 1980s and the devaluation of the Nigerian currency led to an economic crisis and a re-evaluation of policies. As a result, imports of wheat were banned from 1987 in order to save foreign currency. This did nothing, though, to lessen the people's demand for bread (Fig. 11.12). The government's response was to encourage an expansion of home production by increasing ninefold the price paid to wheat farmers between 1985 and 1990. This incentive caused wheat production to increase from 15 000 tonnes in 1983 to 140 000 tonnes in 1988-9.

?

6 Draw a flow diagram of events that led to Nigeria's wheat trap.

7 Evaluate the effects of donating wheat as food aid to Nigeria.

India's Operation Flood

Figure 11.13 India

Figure 11.14 Farmers queue to deliver milk to Operation Flood

India's milk scarcity

India (Fig. 11.13) has 250 million cattle, but their milk production is low. This is because over the centuries cattle have been bred for draught purposes (to pull carts and ploughs) rather than for milk. In 1965, in an attempt to boost rural milk production, India's National Dairy Development Board was founded.

Europe's skimmed milk

Operation Flood (OF), introduced in 1970, was a scheme to import skimmed milk powder donated by the EC. The idea was to make up the skimmed milk powder with water and sell it to urban consumers. The money generated from the sales would then be invested to increase Indian milk production. The scheme had two major advantages: it allowed the EC to rid itself of embarrassing surplus stocks of milk powder; and it helped India to develop its own dairy industry.

The OF system

OF became involved with farmers, encouraging them to sell their surplus milk to their local village dairy co-operative (Fig. 11.14). The village dairy co-operatives usually receive milk twice daily. The milk is collected by truck and taken to a local chilling plant or dairy where it is processed before being delivered to urban consumers. Farmers are paid weekly, which provides poor rural producers with a steady cash flow. In addition, OF guarantees to take all of their production and also provides veterinary advice and fodder.

As well as supplying farmers with a regular income, OF has tackled the problem of seasonality in milk production. While demand for milk remains fairly constant throughout the year, the supply tends to be

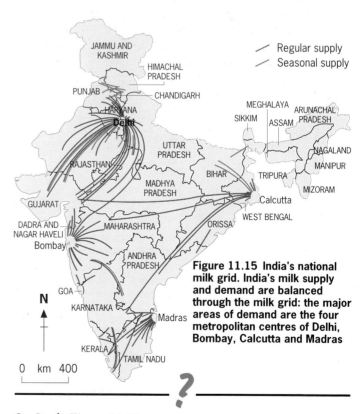

Figure 11.15 India's national milk grid. India's milk supply and demand are balanced through the milk grid: the major areas of demand are the four metropolitan centres of Delhi, Bombay, Calcutta and Madras

8 Study Figure 11.15.
a Where are the main milk surplus regions in India?
b Where is most of the milk supply for the four metropolitan centres likely to come from?
c Which metropolitan centre imports milk over the longest average distance?
d Suggest what factors might influence the geographical pattern of milk movement in Figure 11.15.

seasonal. To bridge this gap, surplus milk produced in the spring is converted to milk powder to make up the deficit in the lean months. Meanwhile, problems of regional surpluses and shortages have been dealt with by the setting-up of a National Milk Grid (Fig. 11.15).

Results

By 1994, 450 towns and cities were supplied with milk from around 50 000 dairy co-operatives scattered all over India. The money from the sale of dairy food aid has been spent on factories, transport and the rest of the infrastructure needed to collect and distribute milk quickly and efficiently. Today OF is the world's largest dairy development scheme.

Criticism

Operation Flood is a controversial aid project. Its supporters call it 'the intelligent use of food aid'. However, it has many critics with four major areas of concern:

1   Some critics say that OF helps only those farmers who have surplus milk production. As these are the better-off farmers, OF is unlikely to reduce poverty. However, a survey in 1984 showed that 72 per cent of co-operative dairy members were either small farmers or agricultural labourers, and that many were from underprivileged castes or tribes. Thus, although it is true that the poor have fewer cows and produce less milk, they do contribute to and benefit from OF.

2   Farmers have been encouraged to sell all of their production to OF, thus depriving their families of valuable milk protein. However, the counter

argument is that OF members can buy cheaper calories with the cash they earn.

3   Some critics argue that OF increases dependence on food aid, foreign equipment and foreign expertise. In fact, in 1987 imported stocks of milk powder (SMP) in milk equivalent was only 6 per cent of the total milk production of the Indian dairy industry.

4   OF has increased dependence on exotic cattle breeds (e.g. Jerseys, Friesians) imported from economically developed countries. Cross-bred cows yield more milk than traditional Indian breeds, but suffer problems of heat stress, and their male offspring are of little use for ploughing. Furthermore, their milk is not as rich as many Indians prefer; they also require more intensive feeding and are more susceptible to disease. Attempts to breed cross-strains with advantages of both may help counter this.

Conclusion

Overall we can say that OF is well intentioned and a qualified success. Similar schemes have been started for fishing, fruit and vegetables and for oilseed using OF as a model. Other economically developing countries are now also interested in the OF experience for their own agricultural development.

9   Suggest why OF has been described as 'the intelligent use of food aid'. Do you agree that OF has been a 'qualified success'? Explain your answer.

11.4 Modern technology and food production

While food aid is appropriate for short-term emergency relief, it is not a long-term solution to the problem of food supply for centres at risk. A more satisfactory response in the long run is to increase local food production. Technology can play a key role in this. Without doubt, the most significant technological advance in food production in the last 35 years has been the Green Revolution.

The Green Revolution

The Green Revolution was the result of an intensive plant breeding programme which produced new, high-yielding varieties (HYVs) of wheat and rice (Fig. 11.16). The new wheat varieties were bred in Mexico and the new rice varieties in the Philippines. These HYVs quickly spread to many parts of the developing world, with some spectacular results. For example, between 1967 and 1992 the world's rice harvest

doubled, and there was a similar upturn in wheat production.

Although much of the research and development money and expertise came from economically developed countries, the Green Revolution is unusual because it is an example of technological transfer *within* the developing world.

Green Revolution

Figure 11.16 Examining disease resistance characteristics of rice varieties, IRRI, Philippines

History

The original objective of the research was to find varieties which:

1 were 'day length-tolerant' i.e. varieties that could grow at different latitudes and altitudes;

2 were short stemmed, to reduce the problem of the plant being blown over by wind;

3 could respond well to high moisture and nutrient status.

Wheat

The story began in 1943 when the Rockefeller Foundation, in partnership with the Mexican government, set up a research project in northern Mexico to improve the local variety of wheat. The initial results were spectacular and received a great deal of publicity. The seeds of the improved wheat varieties were released to the rest of the world in the early 1960s.

Rice

In 1960 the International Rice Research Institute (IRRI) was founded at Los Baños in the Philippines with the financial support of the Rockefeller and Ford Foundations. Part of the IRRI's job was to collect varieties of rice from all over the world to develop a seed bank for possible use in breeding. By 1966 the IRRI was able to release its first so-called 'miracle' rice, IR8. The new rice has several advantages over traditional varieties (TVs) (Table 11.5).

Table 11.5 Advantages of modern rice varieties

- HYVs have shorter stems and narrower leaves than TVs. This increases the plant's strength and prevents 'lodging' (falling over, weakening the plant). There is also an increase in the ratio of the weight of useful grain to that of the rest of the plant (which has less value).
- The rice plants have a standard height, which reduces the problem of mutual shading.
- Unlike traditional varieties, the new rice is insensitive to photoperiod or day-length. This allows up to three crops to be grown on the same plot each year.
- The new rice matures rapidly. For instance, IR8 matures in 120 days and IR28 in just 105 days. This compares with about 160 days for TVs.
- Yields from the new rice are higher than TVs. However, improved yields depend on a package of inputs, including chemical fertilisers (especially nitrogen), pesticides and herbicides, and irrigation water.

By 1988 there were 500 HYVs of rice, of which 150 were bred by the IRRI. IR64 is the highest-yielding and most resistant to pests and diseases. It took over five years and 18 000 breeding experiments to perfect. HYVs of wheat and rice have spread quickly (Table 11.6) and by the end of the twentieth century will have surpassed the production of TVs.

Table 11.6 The spread of HYVs (% output) (*Source*: Alexandratos, 1988)

	1982–4	2000 (projected)
Sub-Saharan Africa	10	30
North Africa/Middle East	40	70
Asia (excluding China)	35	60
Latin America	40	45
93 other economically developing countries	34	57

Revolution or evolution?

It is wrong to see the Green Revolution as a single event. Since 1960 it has gone through several distinct phases, or areas of research.

Phase 1

The first phase, in the 1960s, was a period of great optimism. The 'miracle' varieties of wheat and rice spread rapidly and had obvious benefits for peasant farmers. This early phase saw an increase in output at twice the rate of population growth; a reduction in annual harvest fluctuations; and lower food prices for consumers. Also, the higher yields from HYVs meant that farmers could reduce the area of land used for cereals and so grow a wider range of crops. Often these additional crops were grown for cash. Finally, HYVs required a higher labour input than TVs to help with weeding, fertiliser and water control. This benefited farm labourers seeking employment.

Phase 2

By the early 1970s, enthusiasm for the Green Revolution had decreased. The third of the initial objectives – to find varieties which could respond well to high moisture and nutrient status – became the most important factor for research, while the second objective – to contain disease – was replaced by the use of pesticides. As a result, it became obvious that the Green Revolution was not just about HYVs of wheat and rice, but a total 'package' including irrigation and **agrochemicals**. Smallholders were therefore placed at a disadvantage, compared to their larger and more prosperous neighbours, for taking up this cropping innovation (see Sections 7.10–7.11). Not only did they have little cash or credit to invest, but they were also less well informed, less able to take risks, and had only limited access to irrigation water. In theory, the package should have helped everyone. But in practice, it benefited the larger, more prosperous farmers, and widened the gulf between rich and poor.

The Green Revolution also had a political dimension. It was planned as a technological short-cut to development which would avoid major social or political upheaval. It therefore reduced pressure to change the existing social order which in many economically developing countries was, and still is, unjust and oppressive. For this reason, the Mexican government supported the original breeding project, hoping that it would avoid the need for land reform. Other problems surfaced in the early 1970s. HYVs, unlike TVs, had little immunity to pests and diseases so more pesticides were needed. The chemicals from these often remained in the soil, causing damage to the environment.

HYVs became geographically concentrated in countries and regions with similar environmental conditions to those where the new varieties were bred (north Mexico for wheat and the Philippines for rice) and which had ready access to irrigation, fertilisers and pesticides. Africa, though, as the hungriest continent, gained little benefit. Conversely, in India, the irrigated areas of the Punjab were well suited to HYV wheat growing, while HYV rice cultivation thrived in Tamil Nadu. Moreover, as both regions were already fairly prosperous, their success in the Green Revolution widened regional disparities of wealth across India. Meanwhile, the less-favoured areas suffered because the price of their wheat and rice was undercut by the HYVs. None the less, there was some compensation. Because HYV cultivation demands greater inputs of labour, the Green Revolution created new jobs and encouraged some labour migration to the more prosperous areas.

Phase 3

In the later 1970s, evidence emerged that smaller farmers *were* adopting HYVs. Indeed, in some areas small farmers planted the greater part of their cropland with HYVs. At the same time, HYVs allowed more prosperous farmers to reserve some of their rice land for more tasty varieties such as *basmati* or for other food crops. The general view that the poor were losing out in absolute terms was no longer acceptable. Instead, living standards among the poor were slowly improving, even though, compared to their richer neighbours, they were relatively poorer.

Phase 4

In the 1980s there was a return to optimism. The Green Revolution scientists realised that not enough had been done to help disadvantaged farmers such as those in remote or non irrigated areas, those growing poor quality **subsistence** crops, and those who were landless. The landless are particularly significant, because by 1995 they will outnumber farmers among the rural poor in the developing world.

Two major trends became apparent during this phase. First, there was an interest in developing HYVs in coarse grains like millet, sorghum and lentils.

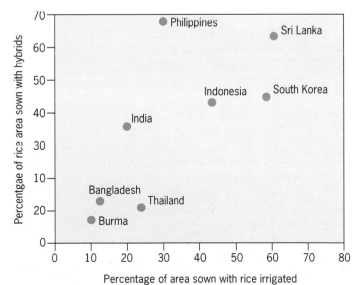

Figure 11.17 High-yielding varieties of rice and irrigation in selected countries (*Source:* Grigg, 1984)

?

10a Study Figure 11.17. What does it tell you about the success of introducing HYVs of rice in economically developing countries?

b On the evidence of Figure 11.17 would you say that HYVs are the answer to food shortages in the economically developing world? Explain your response.

Green Revolution

Traditionally these cereals have been regarded as 'poor people's food'. For instance, in Asia rice and wheat account for 77 per cent of grain production; in South America 24 per cent; and in Africa only 10 per cent (Dixon, 1990). There was also an attempt to identify the needs of semi-arid and other environmentally less-favoured regions which depended on rain-fed agriculture. Second, there was a shift away from the goal of yield increase to other qualities such as disease resistance, and the response to moisture shortage.

Phase 5

In the 1990s we are beginning to see the impact of biotechnology, where micro-biological techniques are used to modify crop characteristics. Through this, genes from plants, animals and micro-organisms become the raw materials for the commercial development of new agricultural products. There has been a shift from the Green Revolution to the Gene Revolution.

?

11 As an agronomist and adviser to the FAO write a report for your employer which:
a describes and explains the current distribution of HYVs of wheat and rice.
b assesses the potential for expansion in Asia, Africa and South America of HYVs of wheat and rice.
c sets out a case for improving varieties of coarse grain crops.
d sets out a case for expanding rain-fed crops and cropping techniques.
Base your report on Figures 11.18 and 11.19 and other information in this chapter.

12 Study Figure 11.20. Assess the negative aspects of the Green Revolution and suggest how they might be reduced.

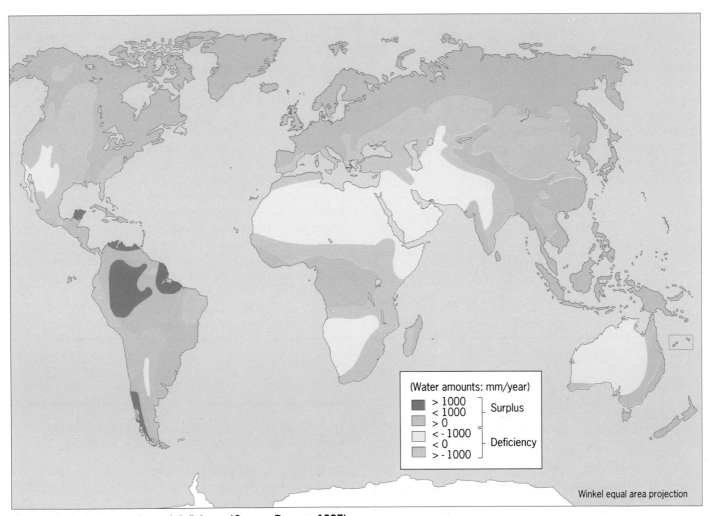

Figure 11.18 Water surplus and deficiency (*Source*: Barrow, 1987)

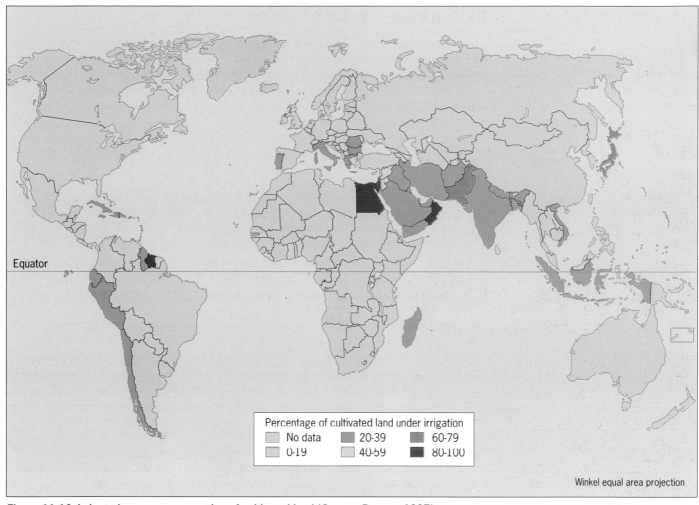

Figure 11.19 Irrigated area as a proportion of cultivated land (*Source:* Barrow, 1987)

Percentage of cultivated land under irrigation

No data	20-39	60-79
0-19	40-59	80-100

Winkel equal area projection

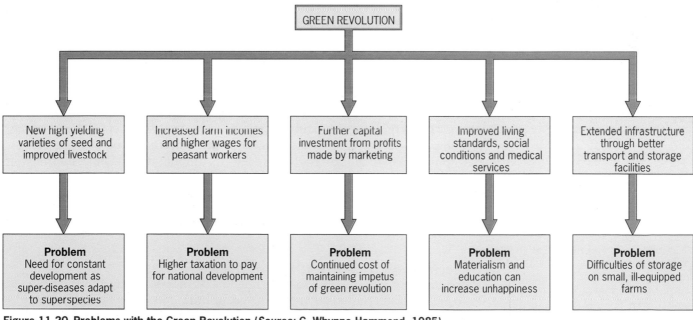

GREEN REVOLUTION

New high yielding varieties of seed and improved livestock	Increased farm incomes and higher wages for peasant workers	Further capital investment from profits made by marketing	Improved living standards, social conditions and medical services	Extended infrastructure through better transport and storage facilities
Problem Need for constant development as super-diseases adapt to superspecies	**Problem** Higher taxation to pay for national development	**Problem** Continued cost of maintaining impetus of green revolution	**Problem** Materialism and education can increase unhappiness	**Problem** Difficulties of storage on small, ill-equipped farms

Figure 11.20 Problems with the Green Revolution (*Source:* C. Whynne-Hammond, 1985)

11.5 Biotechnology

Biotechnology means the application of modern, laboratory-developed, high technology in food production. Whereas the Green Revolution involved introducing new varieties of primarily wheat and rice in selected areas, biotechnologies have the potential to enhance all agricultural production. There has been an explosion of interest in biotechnology since the 1980s and this has affected both livestock and crops.

Livestock

Animal breeding to improve milk yields or carcass size is comparable to plant breeding which led to the Green Revolution. Artificial insemination using frozen sperm has become a standard technology since the 1960s. More recently, embryo transfer (ET) has appeared. Ova (eggs) are harvested from high-quality cows, fertilised in test tubes, and implanted into a foster mother (Fig. 11.21). Although ET is costly, it is economically worthwhile. For example, it is possible to implant Aberdeen Angus eggs into a Friesian cow, thus stimulating milk production and producing high-quality beef as well.

The use of chemicals is more worrying. In the past, some antibiotics used to treat infections in humans were given to livestock to make them grow faster. As a result some bacteria became resistant to the drugs, which endangered farm workers and their families. Since 1968 there has been a farm ban on antibiotics which are useful to humans, and no animal on antibiotics can be slaughtered without a month's 'holding period'.

Bovine somatotropin (BST) is a genetically engineered growth hormone that can be used in dairy cattle. It increases milk yields by up to 25 per cent for short periods, but has unknown effects on humans. In 1989 the EC imposed a ban on the use of BST, which was later extended to 1993. For a period, the UK government permitted its use on selected farms and let the milk from these experimental animals into the national supply.

Figure 11.21 Dexter calves, born after frozen embryos were flown to Western Australia from Canada and surgically implanted into the mother cows.

Crops

There are several approaches to the use of biotechnology in crops. Tissue culture involves growing small pieces of plant tissue or individual cells in culture. These laboratory conditions encourage the growth of thousands of identical plants and provide an efficient way of taking cuttings from a single plant. This is the basis of plant cloning, or micropropagation of plants which doubles the speed of traditional plant breeding. Another process is genetic manipulation, which is a relatively recent technique. It consists essentially of mixing and matching genes from unrelated species. This method will provide plants and animals with properties which they would not develop naturally. So far, only a few crops have been genetically manipulated. New genes are carried into plant cells either by injection of DNA with a very fine needle or by microbes.

Biotechnology: opportunity or obstacle?

Biotechnology research has so far tended to be mostly centred in the industrial world. There is therefore a risk of failing to focus on the needs or interests of poorer farmers in less developed areas. Consequently, although biotechnology *could* be used to increase food production, it could also hinder development or create serious hardship for rural communities (Table 11.7).

?

13 Read Figure 11.22. Construct a table setting-out the values and beliefs of those who favour gene technology and those who are against. Clarify your own view on this issue by identifying the values and beliefs you hold.

STUART WAVELL'S
·PEOPLE·
Ripe for research?

Ian Crawford

Easy commercial pickings: Don Grierson's work has resulted in several tomato patent applications. 'But I don't think we've crossed the Rubicon,' he says

It's firm. It's sweet. It's long lasting. It's the super tomato. The dream of Professor Don Grierson and his team of genetic researchers at the University of Nottingham ripened visibly last week with the injection of a £424 000 government grant, topped off by a £200 000 dollop from the European Community.

Does it deserve a raspberry? Yes, chorus the old order of biochemists, whose article of faith is that nothing in nature is safe unless it has proven credentials of survivability. No, reply their usurpers, the new bankable breed of biogeneticists, who see nothing amiss to putting numbers to biology and dialling up – in the case of the perfect tomato – a more than doubled shelf life of 10 days.

Grierson, professor of plant physiology in the university's department of environmental science, has won an international reputation for his work over the past 16 years into plant genes of which the quest for an improved tomato is a short-term spin-off.

Broadly speaking, plants have 50 000 genes, about which very little is known, as Grierson readily admits. Gene technology allows him to identify one gene, isolate it in a test tube, and move it from one plant to another. This, he says, is simply a more refined process than the random technique of plant breeding, where genes are reshuffled through the transfer of pollen.

He holds that even when a gene from a bacterium is inserted into a tomato, a process known as transgenics, the plant remains quintessentially a tomato.

But his new germ plasm would not be habitat-derived, nor would commercial considerations allow its safety to be tested over generations, his critics argue. By adding a gene and scrambling the controls, he might introduce a Trojan horse for which no laboratory test exists. Isn't he playing genetic roulette?

'It's not roulette, because you can't do these experiments without a scientific understanding of what's going on,' he said 'Roulette implies a degree of randomness that's misleading.'

'I think it's wrong to claim that at one stroke one could alter the properties of plants so that they are resistant to all fungi, all insects or all viruses. What we are dealing with is small, incremental improvements. It's wrong to give the impression that something dramatic and awful is about to happen.'

Just the thin end of the wedge? Grierson replied to the effect that understanding of the environment was necessary to identify its pitfalls and generate improvements. Disappointingly, he cited the benefits of antibiotics. 'Also, we are living in a democracy and it's up to everybody to express an opinion. They can make choices.'

Whose choices, after the genie is out of the flask? As the Prince of Wales noted last week, tomatoes were first developed by indigenous cultures in the Amazon and Andes. Won't his all-purpose tomato, the answer to every housewife's prayer, hazard the survival of precarious wild stocks as well as Third World exporters who are sustaining the species' diversity?

Grierson thought the loss of wild strains would be 'unfortunate' but unlikely, for commercial reasons to do with the vulnerability of cultivated crops. 'Maybe we won't need certain wild species for 50 years or more,' he said, a shade optimistically. 'But we should make sure, now, that we retain them.'

In his rain forest appeal last week, the Prince of Wales said of forest products 'We spend millions of pounds on 'improving' these foods — trying to make them sweeter, more colourful, or tastier. Perhaps that investment might be better applied to pursuing new products from the forests?'

But genetic engineering is attractive to investors because it promises a relatively quick return, and the product can be *patented*. Grierson said that his university, in conjunction with ICI, had applied for 'several' patents on tomato improvements, as had 'universities in the UK, Europe and America, and many companies throughout the world'.

However, Grierson expressed doubts on this score. Current thinking was that the discovery of a gene was insufficient – some added value was required. 'It remains to be seen what is going to happen when we come out of the tunnel,' he added.

We have already emerged, according to Bill McKibben in his recent book The End of Nature. 'It is the simple act of creating new forms of life that changes the world — that puts us forever in the deity business,' he writes.

Grierson found that too simplistic a view. 'I do agree that the intensification of our activities in agriculture and industry has brought tremendous improvements as well as some detrimental effects. But I don't think we've crossed the Rubicon. I think we've been going through a long, slow learning curve.'

But we haven't learned to slow down. In a world baffled by Aids, does his work merit the risk? He replied that his team's experiments were licensed, scrutinised, policed, debated and contained. 'There's a small risk when I cross the road. I think that the kind of gene manipulation that we'll be doing at Nottingham is a much smaller risk than that.'

Figure 11.22 Gene technology (*Source: The Sunday Times*, 11 Feb. 1990) © Times Newspapers Ltd 1990

Table 11.7 Biotechnology and social and economic issues

Risk	Impact
Substitution	Biotechnology threatens to eliminate or displace traditional agricultural products e.g. current research focuses on substitutes for tropical oils and fats. In some cases, economically developing countries may rely on these goods as export commodities and as a source of foreign exchange.
Genetic diversity	**Biodiversity** is threatened as traditional crops and livestock breeds are diluted by imported breeds through cloning or ET e.g. automated nurseries in Chile can propagate up to 10 million eucalyptus seedlings – all identical clones.
Biosafety	There is a danger in introducing genetically engineered plants into centres of diversity e.g. resistant plants may release genes to weeds growing close by. The resultant herbicide-tolerant weed could be difficult to control, harming future crop production as well as the surrounding ecosystem.

Summary

- Food security rather than food availability is the key to understanding famine and hunger. The food security of individuals is threatened when they lose their entitlements.

- It is estimated that the world could adequately feed a population of at least 8 billion if farming resources were fully utilised.

- There are two common approaches to solving problems of food shortage: food aid and technological change to improve food output. Food aid is a short-term solution; technological change has more lasting effects.

- Food aid is given for a variety of reasons other than purely humanitarian. It does not necessarily go to the most needy countries.

- Food aid can have an adverse impact on traditional agriculture in recipient countries, and may lead to dependence on imported food.

- The most effective food aid is that which helps a country to develop its food production potential e.g. Operation Flood in India.

- The Green Revolution, based on new high-yielding varieties of cereals, has had spectacular success in increasing wheat and rice production in the economically developing world.

- The benefits of the Green Revolution have been spatially and socially uneven.

- Responses to the Green Revolution have gone through several phases of optimism and pessimism since the early 1960s. Research is currently addressing the problems of small farmers, dependent on coarse grains (e.g. millet, sorghum) and rain-fed agriculture.

- Biotechnology has enormous potential to increase world food production but raises many difficult issues.

12 Agricultural policy and food supply

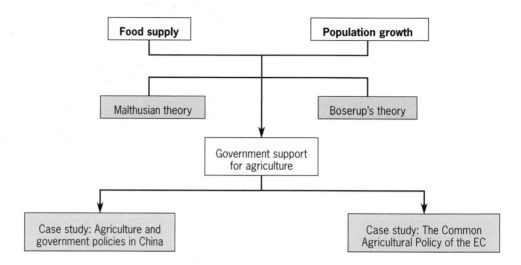

```
┌─────────────┐              ┌──────────────────┐
│ Food supply │              │ Population growth │
└─────────────┘              └──────────────────┘
        │                             │
   ┌─────────────────┐         ┌────────────────┐
   │ Malthusian theory │       │ Boserup's theory │
   └─────────────────┘         └────────────────┘
                    │
           ┌──────────────────┐
           │ Government support │
           │ for agriculture    │
           └──────────────────┘
         │                              │
┌──────────────────────┐      ┌──────────────────────┐
│ Case study: Agriculture and │ │ Case study: The Common │
│ government policies in China │ │ Agricultural Policy of the EC │
└──────────────────────┘      └──────────────────────┘
```

12.1 Introduction

We have seen in Chapter 11 how problems of food supply in economically developing countries are being tackled through food aid and the introduction of modern technology. Yet in spite of food shortages, governments in these countries have tended not to make agricultural reform the main thrust of their policy, favouring more prestigious activities such as manufacturing industry. Ironically, in the developed world, where food surpluses rather than food shortages are likely to be the main problem, agriculture and food production are often more tightly controlled by government policies than any other sector of the economy.

In this chapter we describe government policies in China and the EC; the reasoning behind them; and their impact. We start, however, by considering the relationship between food supplies and population growth.

12.2 Food supplies and population growth

There are two opposing viewpoints concerning the relationship between food supplies and population growth: the pessimistic view first put forward by Thomas Malthus; and the optimistic view often associated with Esther Boserup.

Malthusian theory

In Chapter 10 we argued that famine and hunger frequently result from a complex interaction between social, economic and political factors. Thomas Malthus, an eighteenth-century English clergyman, took an altogether simpler view, arguing that population pressure, above all else, was the direct cause of famine.

Malthus first set out his ideas in his classic text *An essay on the principle of population*, in 1798. Malthus assumed that land was an unchanging factor

Malthusian theory

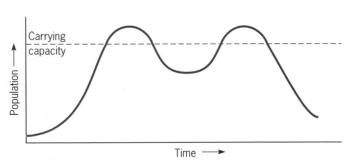

Figure 12.2 Population change and resource limits

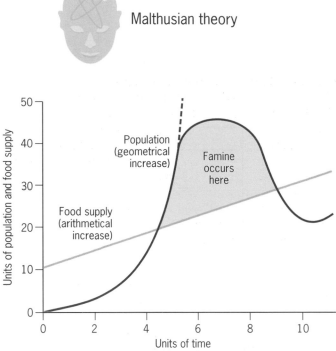

Figure 12.1 Illustration of Malthus' view on population growth and food supply

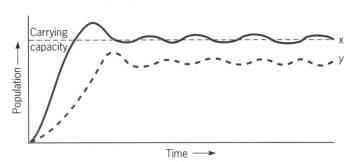

Figure 12.3 Population change and resource limits

and that the food supply can only be increased at a steady (arithmetical) rate (Fig. 12.1). Population, however, increases geometrically and will inevitably outstrip the growth of food supplies. Furthermore he argued that people have to control their fertility, otherwise they will experience falling living standards, disease, famine and starvation. This, though, will bring the population down to supportable levels again.

Malthus's 'doom and gloom' prophecy proved inaccurate, at least for Britain in the nineteenth century. Although the population of England and Wales soared from 9 million in 1800 to 32 million in 1900, food shortages did not occur. The reasons for this were improved farming technology which boosted crop yields, and the unprecedented growth of international trade in foodstuffs. Thus Britain came to rely on food imports such as grain from North America, beef from Argentina, butter from New Zealand, and tea from India. Indeed, cheap food became a major policy of successive British governments, from the repeal of the Corn Laws in the 1840s to joining the EC in 1973.

Even so, Malthusian ideas have remained popular among some economists (see Table 12.2). These 'neo-Malthusians' argue that current food shortages in some economically developing countries are the result of overpopulation. In other words, there are too many people for the resources available to feed them at current levels of technology. Hence they argue that the problem of food shortage will only be solved when population growth is controlled.

Carrying capacity

We can define carrying capacity as the maximum sustainable population in an area which can be supported at a certain standard of living. According to Malthus, population would overshoot the carrying capacity until starvation and fewer births caused the population to fall, thus bringing it into balance with the carrying capacity. This pattern is common in animal populations.

?

1 Study the relationship between population growth and carrying capacity in Figures 12.2 and 12.3. Suggest what events or government policies could produce:
a the curve in Figure 12.2.
b curve x in Figure 12.3.
c curve y in Figure 12.3.

Boserup's theory

In 1965 Esther Boserup presented an alternative view in her book *The conditions of agricultural growth*. Although she accepted that population growth would increase the demand for food, she argued that it would not necessarily lead to food shortages. Instead, food scarcity would push up prices and give farmers an incentive to increase output. This could be achieved in several ways: by working longer hours; by cultivating more land; by using more advanced technology; or by intensifying production (for example by irrigation or multi-cropping).

Boserup proposed a simple five-stage progression in which each stage represents a significant increase in both the intensity of the cultivation system and the number of families it can support (Table 12.1).

Table 12.1 Boserup's five-stage progression

Stage 1	Forest-fallow cultivation	Consists of 20–25 years of letting fields lie fallow after 1 or 2 years of cultivation.
Stage 2	Bush-fallow cultivation	Involves cultivation for 2 to as many as 8 years, followed by 6–10 years of letting lands lie fallow.
Stage 3	Short-fallow cultivation	There are 1–2 years when the land is fallow and only wild grasses invade the recently cultivated fields.
Stage 4	Annual cropping	The land is left fallow for several months between the harvesting of one crop and the planting of the next. This stage includes systems of annual rotation in which one or more of the successive crops sown is a grass or other fodder crop.
Stage 5	Multi-cropping	This is the most intensive system of agriculture where the same plot produces several crops a year and there is little fallow period.

?

2 Draw a similar graph to Figure 12.2 to illustrate Boserup's theory. (Note that the carrying capacity will increase in stages, as the intensity of cultivation increases.)

3 Suggest what events or policies could produce the curve or curves on your graph.

12.3 Can the growing population feed itself?

?

4 Study the forecast crop yields by 2050 (Fig. 12.4).

a If there is no increase in the harvested area by 2050, and assuming a world population of 10 billion, calculate the percentage increase in yields needed if the world's population is to be fed with an average of: 4000, 6000 and 10 000 kcals/day.

b Make similar calculations to **a** assuming that the world's harvested area increases by 50 per cent by 2050.

c Comment on the implication of your findings for population–food supply relationships over the next 50 years or so.

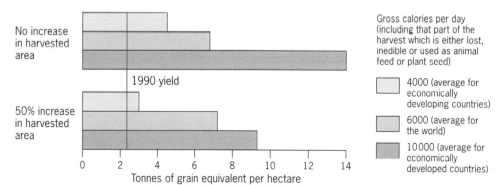

Figure 12.4 Crop yields needed in 2050 to feed a world population of 10 billion (*Source: Bongaarts, 1994*)

The debate on population and food supplies continues, and remains as polarised as ever. Today's pessimists are the environmentalists who predict not only famine but ecological disaster if population growth continues unchecked. Opposing them are the technologists. They believe that the earth could comfortably feed more than twice the current world population without causing widespread environmental damage. The arguments of these two groups are presented in more detail in Table 12.2.

Table 12.2 The population growth and food supply issue: who's right? (*After*: Bongaarts, 1994)

Environmental pessimists	Technological optimists
Attitude 'Human numbers are on a collision course with massive famines . . . If humanity fails to act, nature will end the population explosion for us – in very unpleasant ways – well before 10 billion is reached.' (Ehrlich and Ehrlich; quoted by Bongaarts, 1994)	**Attitude** Technological innovation will raise food production more than enough to feed the world's projected 10 billion population in 2050 without environmental catastrophe.

Environmental pessimists

Beliefs
- The world's population will double (to 10 billion) between 1990 and 2050.
- To feed this massive population growth will mean intensifying farming practices, causing irreparable damage to the environment.
- Although world food output increased by 117 per cent between 1965 and 1990, massive population growth during this period meant only a modest improvement in average per capita food production. In Africa food output per capita actually declined.
- In the long-term, sustainable growth of food production is impossible.
- The number of people with energy deficient diets increased in Africa, Latin America and the Middle East between 1965 and 1990.
- The world is running out of suitable land for cultivation. Soil erosion, salinisation, deforestation and pollution by **agrochemicals** are destroying current and potential farmland.
- The **Green Revolution** brought benefits to a limited number of people, and the growth in yields has slowed down. Production per person peaked at 346 kg in 1984 and is now falling. The spread of HYVs of wheat and rice is limited by a lack of capital for irrigation schemes, a shortage of irrigated land and a shortage of agrochemicals in economically developing countries.

Technological optimists

Beliefs
- Between 1965 and 1990 in economically developing countries, average daily food consumption increased from 2063 kcals to 2495 kcals; per capita protein intake increased from 52 to 61 grams a day. There is no reason why these trends cannot continue.
- Three times as much land as is currently cultivated in 93 economically developing countries (excluding China) could be brought into production. This is an additional 2.1 billion hectares.
- There is scope for increases in multiple cropping in the humid tropics, and higher yields generally in economically developing countries. While crop yields in Europe average 4.2 tonnes/hectare, in Africa they are barely one tonne/hectare.
- Green Revolution technologies could be more widely introduced. Currently only one third of seeds planted in economically developing countries are HYVs.
- Huge quantities of food are wasted. Major losses occur during distribution and storage. Currently humans consume only 60 per cent of all harvested crops, with 25 to 30 per cent lost before they reach the consumer (NB: FAO's estimates for losses are lower: 6 per cent for cereals, 11 per cent for root crops, 5 per cent for pulses).
- For many years world food prices have been falling, owing to over-production in the EC and North America. Cereal prices fell by one-third on international markets between 1980 and 1989. Thus people in economically developing countries are in a better position to afford to buy food.

?

5 Study the conflicting views in Table 12.2. (You may notice that the two groups often use the same evidence but interpret it in different ways.)

a Look back at previous chapters in this book to find out more on population growth, food supplies, and the environmental impact of agriculture.

b Clarify your own attitudes and beliefs concerning the issue of population growth and food supplies. You are a journalist writing for a current affairs magazine. Present your views, supported by evidence from Tables 12.3 and 12.4 and other sections of this book. Make your article lively, interesting and informative.

Table 12.3 World population, millions (*Source*: Philips Digest, 1992–3)

	1950	1960	1970	1980	1990	2000 (estimated)
World	2516	3020	3698	4448	5292	6261
Africa	222	279	362	477	642	867
Asia	1377	1668	2102	2583	3113	3713
Europe	393	425	460	484	498	510
North America	220	270	321	374	427	479
Oceana	13	16	19	23	26	30
South America	112	147	191	241	297	345

Table 12.4 World food production, 1965–85 (millions of metric tonnes)

Year	1965	1970	1975	1980	1985
Cereals					
Developing	470	587	683	771	928
Developed	536	618	689	767	919
Root Crops					
Developing	246	302	330	354	371
Developed	243	259	223	185	217
Meat, milk and fish					
Developing	115	136	152	184	224
Developed	387	421	454	489	524
Oil crops, pulses, vegetables, fruits					
Developing	284	322	374	435	505
Developed	230	261	280	304	316
Total	2511	2907	3185	3519	4004

Table 12.5 UK government subsidies, 1990 (*Source*: Marks, 1992)

Industry	Subsidy (£ million)
Agriculture, forestry and fishing	1633
Fuel and energy	196
Mining, minerals, manufacturing	150
Transport and communications	1083
Other economic activities and services	1204

12.4 Governments and agriculture

Governments play an increasing role in supporting and guiding the fortunes of agriculture. Today most countries have some sort of agricultural policy, and agriculture is heavily subsidised throughout the developed world (Fig. 12.5). Often these policies have an influence which greatly exceeds the economic importance of agriculture. For example, in the UK agriculture employs only 2.5 per cent of the working population and creates just 1.25 per cent of the national wealth – yet it has its own government department. No other industry has this status, nor enjoys such a high level of government financial support (Table 12.5). Why is this?

There are three main reasons why politicians give favoured treatment to agriculture:

1   To maintain agricultural incomes at an acceptable level which keeps people on the land and prevents the decline of rural communities.

2   To keep food prices for consumers as low as possible.

3   To achieve national self-sufficiency in basic foodstuffs.

The UK and Japan are the world's largest food importers. However, successive British governments, fearing food shortages in time of war and needing to reduce the trade deficit, have favoured policies which boost domestic food production. The success of these is remarkable: before 1914 the UK imported about 67 per cent of its food requirements but by 1992 this had fallen to 43.5 per cent overall, and to 27.9 per cent for temperate crops.

?

6   Suggest reasons why governments wish their countries to be as self-sufficient in food as possible.

7   In groups, brainstorm what might happen to the areas listed in Figures 12.6–12.11 if subsidies were removed. List possible consequences for: • the environment, • the population, • the government.

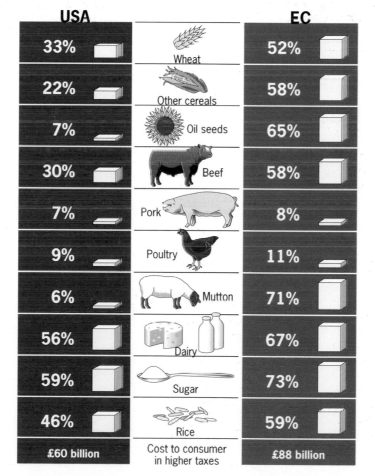

Figure 12.5 Farm subsidies in the USA and EC, 1992 (*Source: The Independent*)

Figure 12.6 Agricultural subsidies in the UK

Regional aid
The poorer farming regions (less Favoured Areas) account for 53 per cent of the total agricultural area. There are two types of Less Favoured Area: • Severely Disadvantaged Areas and • Disadvantaged Areas. In both areas farmers qualify for Hill Livestock Compensatory Allowances (HCLAs) which are payments made for each head of livestock. The aim is to prevent the decline of farming communities and conserve the traditional farmed upland landscape.

Figure 12.7 Less Favoured Area, North Yorkshire

Farm Woodland
In 1988 the Farm Woodland Scheme was introduced for the retirement of agricultural land for long periods. Payments were £190 per hectare, and one-third of the planting must be broadleaved species. In 1992 grants were increased to £250 per hectare for 10-15 years.

Capital grants
Grants are available for expensive improvements such as land drainage and flood defences.

Figure 12.8 Preparing to lay field drainage pipes, Hampshire

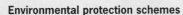

Examples of subsidy

Figure 12.11 Newly planted copse on former farmland, Berkshire

Environmental protection schemes
These include the Countryside Premium Scheme and the Farm and Conservation Grant Scheme. Grants encourage hedge and tree planting, and the protection of Sites of Special Scientific Interest on farmland. Farmers are encouraged to be more environmentally conscious by the Farming and Wild-life Advisory Group. Recent schemes include the establishment of Environmentally Sensitive Areas in 1986 (see Section 6.11) and Nitrate Sensitive areas in 1990 (see Section 3.9). In 1994 a programme to enhance wildlife habitats was launched. For example, farmers would be paid to allow arable land to revert to salt marsh, and to leave 20 metre wide strips alongside river banks free from agrochemicals.

Grants for diversification
Also in 1988, farmers were offered grants to encourage diversification into enterprises which would not add to EC food surpluses. Many forms of diversification have been proposed, from keeping llamas to building golf courses.

Figure 12.9 Grass margin strip between hedgerow and wheat crop, Hampshire

Figure 12.10 Diversification: new golf course on former farmland, Powys

Agriculture and government policies in China

China has a huge agricultural sector (Fig. 12.12). Two-thirds of the working population is employed in agriculture, which accounts for one-third of China's GNP. Only the USA has a greater area of arable land.

In the EC farmers are independent decision-makers, free to select the enterprises of their choice. This is because EC government policies influence farmers only indirectly. In China the political framework is quite different. China has a **command economy**. The government owns the major sectors of the economy including agriculture, and its policies determine most aspects of farming and food production.

China became a socialist republic following the Communist revolution in 1949. It was organised on Soviet lines as a one party, **centrally planned**, authoritarian state. However, since the revolution China has adopted a range of policies on agriculture and rural development (Figs 12.13 and 12.14). These policies have been influenced as much by pragmatism as by changing ideology. The main features of government policy since 1950 are shown in Table 12.6.

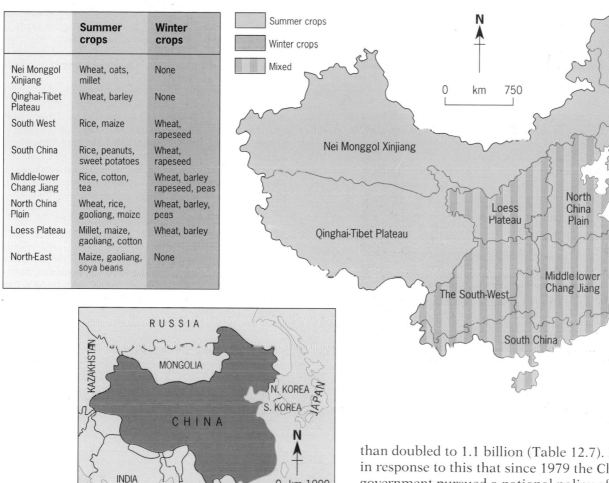

	Summer crops	Winter crops
Nei Monggol Xinjiang	Wheat, oats, millet	None
Qinghai-Tibet Plateau	Wheat, barley	None
South West	Rice, maize	Wheat, rapeseed
South China	Rice, peanuts, sweet potatoes	Wheat, rapeseed
Middle-lower Chang Jiang	Rice, cotton, tea	Wheat, barley rapeseed, peas
North China Plain	Wheat, rice, gaoliang, maize	Wheat, barley, peas
Loess Plateau	Millet, maize, gaoliang, cotton	Wheat, barley
North-East	Maize, gaoliang, soya beans	None

Figure 12.12 China: major agricultural regions (*Source:* C. Pannell, 1982)

Achievements

In 1949 when the Communists took over, the Chinese economy and agriculture were in disorder. Nearly 20 per cent of the world's population lived in the Chinese countryside, and most were poor, malnourished and backward. Famine struck some part of China virtually every year. Since 1950 China's population has more than doubled to 1.1 billion (Table 12.7). It was partly in response to this that since 1979 the Chinese government pursued a national policy of one child per family. The aim is to keep China's population at 1200 million by the year 2000 (Fig. 12.15). So far, such family planning has had considerable success in the cities; in 1984 new born single babies made up 83 per cent of urban births, whereas they only constituted 62 per cent of rural births. Even as its population has grown, and with minimal external assistance, China has been able to feed its people and provide basic needs. Cereal yields are as high as those in Europe and North America, while famine and food shortages have been eliminated.

Agriculture and government in China

Figure 12.13 China under collectivisation: commune workers harvesting wheat

Figure 12.14 China under a liberalised farm policy: mixed crops

?

8 Study Figures 12.13 and 12.14.
a Describe the main differences, looking particularly at field patterns and the range of crops.
b With reference to Table 12.6, explain how government policies might explain landscape change.

9a Using Table 12.7, draw a graph of China's population since 1950.
b Comment on the population trend.
c In 1993, China's population was 1178.5 million. Which of the five projections in Figure 12.15 does the trend fit?

10 Compare China's policies with the theories of Malthus, Boserup and Table 12.2. To which views are these policies closer?

Table 12.6 Changing agricultural policies in China since 1950

1950	Major land reform programme. 40 per cent of China's arable land was redistributed to peasant farmers. However, the government's main priority was industrial reform, supported by massive aid from the USSR.
1952–6	Co-operative movement established, with independent farmers grouping together to form larger buying and marketing units.
1956–7	Collectivisation: all land becomes state-owned. Farms are grouped to run as a unit. Production quotas are set by the government.
1958–9	Introduction of people's communes, with a hierarchy of social and geographical units: villages (teams); groups of villages (brigades); communes. Communes are the basic social unit, providing health care, education and other services, as well as employment. Small-scale rural industries are encouraged to provide inputs (e.g. fertilisers, machinery) to agriculture.
	Peasants are allowed to farm small private plots. In addition to land, draught animals, farm machinery and farm implements become state-owned. Management is at team level; labour is also allocated at this level. Each unit in the hierarchy has production targets set by China's Five Year Plan. In 1960, the USSR withdraws its aid to China. Government priority switches from industry to agriculture.
1966+	The cultural revolution. Rigid central planning with more official/state intervention. The marketing of surpluses for cash, and the production of crops such as fruit and vegetables which could be marketed locally, are banned. Ownership of private plots is criticised. There is a significant fall in marketable outputs leading to food shortages in towns and cities.
1978+	Deng Xiaoping succeeds Mao (died 1976) and introduces a more liberal policy. There is less concern with equality and ideology. The priority is economic growth and the modernisation of agriculture. Government reforms reintroduce family farms and abolish the commune system. Although the land is still under state ownership, the farmer is no longer a collective employee, but an independent farmer. These policies have allowed prices and profits to return to agriculture and have successfully raised output. In future, further liberalisation is likely. China wants larger farms with fewer workers for greater efficiency. It is moving more towards capital intensive, commercial agriculture. These new policies could eventually mean a reduction in the farm population of 35 to 40 per cent.

Table 12.7 China: population (thousands)

1950	1960	1970	1980	1990	2000 (est.)
554 760	657 492	830 675	996 134	1 139 060	1 299 180

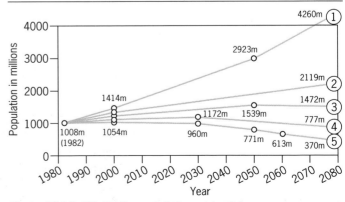

Figure 12.15 China's five population projections

The Common Agricultural Policy of the European Community

At the end of war in 1945, Europe's agriculture was in disarray. During the war, food production on the continent fell and markets were severely disrupted. There were acute food shortages and even famine in parts of the Netherlands and Germany in 1945. It took a long time for food production to recover and rationing remained until well into the 1950s. By 1957, therefore, when the EEC was set up by the Treaty of Rome, governments were eager to increase food output as quickly as possible. Since 1957 the EEC has come a long way. For example, by 1993 more than 60 per cent of the entire EC budget was being spent on agriculture, and support payments averaged £13 700 per English farm.

The stated aims of the Common Agricultural Policy (CAP) were:

- To increase productivity.
- To ensure that farmers, farm workers and their families had a 'fair' standard of living.
- To keep prices for agricultural products steady.
- To maintain regular food supplies for the markets.
- To ensure 'reasonable' food prices for consumers.

Since then, the CAP has undergone considerable change. CAP policies are funded by the European Agricultural Guidance and Guarantee Fund (more often known by its French acronym FEOGA). FEOGA in turn gets its funding from VAT and levies on imports from outside the EC.

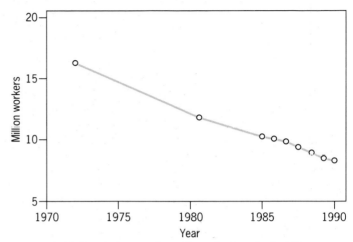

Figure 12.16 Population employed in agriculture in the EC, 1970–90 (*Source:* European Commission)

1 Structural policies: the Guidance Fund

The Guidance Fund seeks to modernise EC farms. In 1968 the Mansholt Plan (which was the preliminary to the CAP) painted a gloomy picture of European agriculture: oversupplied with labour and with too many poor, small and fragmented farms. In fact, there has been a dramatic drift away from land in the EC since 1970 (Fig. 12.16). However, although EC farm policies are only partly responsible for this trend, these structural problems have since been addressed by the Guidance Fund.

11 Compare the origins and aims of CAP with the theories of Malthus, Boserup and Table 12.2. To which view is the CAP closer?

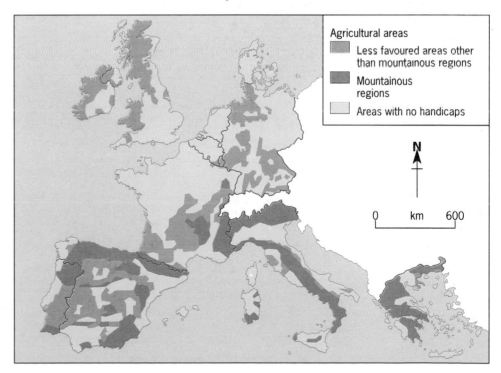

Figure 12.17 Less Favoured Areas in the EC (*After:* European Commission, 1989)

 CAP

Aid has also been directed to specific problem regions in the EC. Thus the Hill Farming and Less Favoured Areas directive (1975) aimed at maintaining farming in marginal upland areas. By the mid–1990s a large portion of EC farmland qualified for such assistance (Fig. 12.17). Since admitting Greece, Spain and Portugal to the EC, regional rural aid has become more important, with investments in irrigation and restructuring vineyards. Consequently, in 1991 regional rural aid took 60 per cent of total expenditure.

2 Price support policies: the Guarantee Section

The Guarantee Section dominates the CAP. In 1993, for example, 91 per cent of CAP expenditure went on price support. Its aims are:

• Common prices for all EC agricultural products. This was finally achieved in 1993.
• To support the principle of EC preference. Essentially the EC is a customs union which aims to promote trade within Europe and minimise imports from outside. There is a tax on imported foodstuffs, though preferential treatment has been given to some developing countries (mostly former colonies), and to New Zealand dairy products, while subsidies occur in the form of export refunds.

Figure 12.18 Farmer ploughing in a Less Favoured Area, Spain

• To support agricultural incomes through manipulating market prices. Many agricultural products receive price support. These support mechanisms are complex and here we give only one example – wheat (Table 12.8).

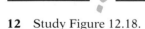

12 Study Figure 12.18.
a What evidence is there that farmers might be disadvantaged in the area shown on the photograph?
b How might the EC's policy help the viability of farming in such an area?

13a Describe the regional variations in agricultural income in the EC in Figure 12.19.
b Describe the regional variations in subsidies per work unit (i.e. for labour) in the EC (Fig. 12.20)
c Compare the distributions in Figures 12.19 and 12.20 and comment on the fairness of the regional pattern of farm subsidies in the EC.

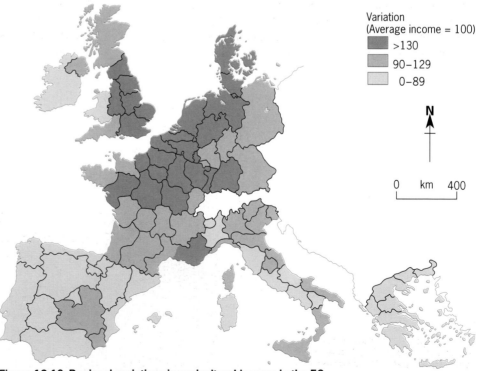

Variation
(Average income = 100)
>130
90–129
0–89

N

0   km   400

Figure 12.19 Regional variations in agricultural income in the EC

Table 12.8 Price support for wheat: three types of prices are set each year

The target price
This is the market price for wheat that farmers will hopefully receive.

The threshold price
This is the minimum price set for imported wheat. It is always set above the target price so as not to undercut EC wheat producers. The difference between the two is called the Variable Import Levy.

The intervention price
This is the minimum market price. It is always set 10–20 per cent below the target price, and is the price at which the EC is obliged to buy.

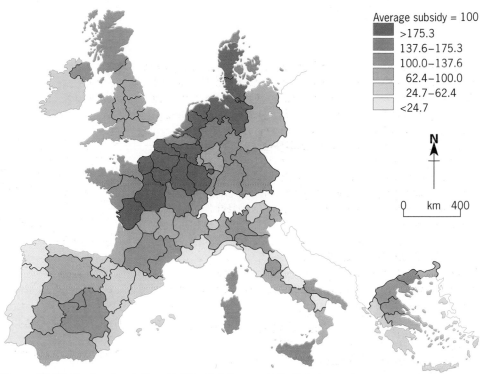

Average subsidy = 100
- \>175.3
- 137.6–175.3
- 100.0–137.6
- 62.4–100.0
- 24.7–62.4
- <24.7

N

0 km 400

Figure 12.20 Regional variations in EC subsidies per unit of work

Food surpluses in the EC: dairy products and cereals

The pricing policies of FEOGA have created self-sufficiency in many agricultural products (Fig. 12.21). This, you may remember, was one of the fundamental aims of the CAP. Unfortunately, the same policies have also resulted in food surpluses which are both expensive to store and dispose of (Figs 12.22–12.23). Although the problem of surpluses is long-standing, attempts to correct the situation have been slow.

Quotas

Dairying was one of the first sectors to receive attention whereby the EC introduced a number of measures to increase demand and reduce production. These include subsidies on the retail price of butter and school milk; food aid in butter oil and skimmed milk to the developing world; a levy imposed on dairy farmers to pay for storing surpluses; and schemes to encourage farmers to switch from dairying to beef production. Milk quotas, introduced in 1984, aimed to

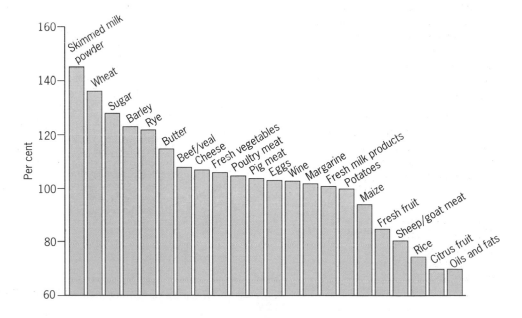

Figure 12.21 EC self-sufficiency in agricultural products, 1990–1 (*Source*: European Commission, 1993)

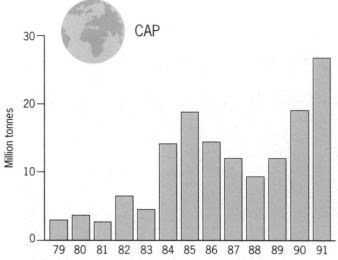

Figure 12.22 Intervention stocks of EC cereals, 1979–91

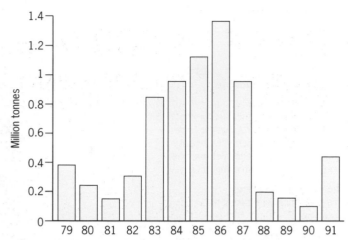

Figure 12.23 Intervention stocks of EC butter, 1979–91

limit output to 1981 levels. Where farmers exceeded their quotas, they had to pay a super-levy of 75 per cent of the value of the surplus. However, in the UK the Milk Marketing Boards operate the quota system, and while farmers producing over their quota pay a levy, all producers share the full penalties for overproduction.

Set-aside

The EC's response to cereal surpluses has been very different. In 1988 it attempted to reduce the area of surplus crops by encouraging **extensification**. One version of this was **set-aside** (Fig. 12.24). Farmers taking part in the original scheme had to retire at least 20 per cent of the farm's arable land for five years. Set-aside land can be left fallow, planted with trees, or used for non-agricultural purposes. Set-aside is a controversial policy (Figs 12.25–12.27). It has been criticised because, in spite of costs, it brings too few environmental benefits to the countryside.

In 1992 set-aside became a part of the latest reform of the CAP which plans by 1995–6 to cut support for cereals by 35 per cent and to abolish support for oilseeds. Instead, farmers will receive a fixed sum per

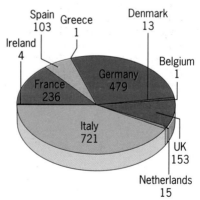

Figure 12.24 EC land set-aside (thousand hectares), 1988–92 (*Source:* European Commission)

?

14 Study Figures 12.25–12.27.
a Draw a matrix assessing the benefits/disbenefits of set-aside.
b Consider the attitudes and values of some of the people affected by set-aside.
c Write an article stating your opinion about set-aside and why the scheme is so controversial.

Set-aside folly

TWO OF the most commonly advanced arguments for farm subsidies are that they maintain employment in rural areas and that they enable farmers to look after the countryside.

These arguments are devastatingly refuted by the Council for the Protection of Rural England's county-by-county survey of farm payments, published today. Lincolnshire, where nearly every bog has been drained, every herb-rich chalk pasture long since ploughed up, receives a staggering £67.6 million a year in subsidy. Of this, £12 million is for set-aside, which pays farmers not to produce anything on 15 per cent of their land. Set-aside must be one of the few subsidies in history which leads directly to unemployment: the farmer has no need of farm workers to cultivate that part of his land.

The absurdity of the Common Agricultural Policy could make a toothless man whistle. Its ever-growing budget, fuelled by corruption, makes another bout of reform urgent. Britain should clarify its minimum demands forthwith. It is to Britain's credit that it has consistently sought a long-term reduction in support for commodities. This it must continue to do. But that in itself will not suffice, for the other 11 countries have yet to show any willingness to reduce subsidies.

With the Gatt agreement signed, and downward pressure on prices, farmers and environmentalists alike are beginning to argue that Britain should seek a change in what farmers are paid to do.

If farmers are to be rewarded for anything, few would dispute that it should be for some public benefit which the market cannot provide, such as wildlife and landscape. It is clear from the CPRE's figures that disproportionate subsidy goes to unattractive prairies — which could make a living if subsidies were taken away — while the preservation of landscape and wildlife receives insufficient support.

To his credit, Mr David Naish, president of the NFU, has recently pledged his support for 'de-coupling' farm payments from commodity support and linking them to countryside management.

Mrs Gillian Shephard, the Agriculture Minister, has been left behind. It is time that she grasped what the public now wants from CAP reform, and used the principle of subsidiarity to bring it about.

Figure 12.25 Newspaper editorial commenting on set-aside (*Source: The Daily Telegraph*, May 1994)

Paid £14,000 to watch the grass grow

Sitting comfrotably: 'You can't conserve the countryside on subsistence farming,' says Suffolk farmer./Photograph by David Mansell

Andy McSmith
Political Correspondent

SINCE 1989, Robert Gosling has been paid more than £14 000 a year to grow nothing but 70 hectares of wild grass. Each weekday morning he drives to an office in Ipswich, where he manages a private estate.

At weekends, he tramps what used to be fields of crops on his Suffolk farm, seeing how the land is slowly recovering from years of intensive agriculture.

It looks a nice way to earn money for nothing. But other farmers see Mr Gosling as 'the rat who left the sinking ship'. He claims: 'They look upon it as the wimpish option, like a form of bankruptcy. They say 'He's into set-aside', rather as they would say: 'He's bankrupt'.'

In 1993, 1500 farmers were paid to grow absolutely nothing. Some received enough to live well without looking for other work. About 10 were paid more than £40 000 a year.

Labour and the Liberal Democrats are fighting this week's European elections on a promise to stop paying farmers to do nothing. Labour's agriculture spokesman, Gavin Strang, calls it 'wholly negative'.

The Council for the Protection of Rural England wants set-aside grants abolished in favour of 'environmental management' payments. Its land-use officer, Paul Wynne, says: 'Instead of going into filling up Range Rovers with petrol, the money should be going into making sure that environmentally sensitive areas are well managed.' But when Mr Gosling describes how

he is managing his land — and he talks with earnest conviction — it makes a crazy kind of economic sense. If he were growing crops, the Government would be compelled, under European law, to subsidise him, which would be more expensive.

Unlike other farmers who have simply allowed the weeds to take over, Mr Gosling is also paid £32 an acre by the Countryside Commission to keep his land in order.

When you take into account the difference betweeen the subsidy he would receive and the cost of paying him to provide a 'public amenity' and habitat for wildlife he calculates the net cost at scarcely £1000 a year.

'You can't conserve the structure of the countryside on a subsistence level of farming. There has to be wealth in the rural areas,' he says.

How that wealth is distributed takes us into the surreal world of Europe's Common Agricultural Policy.

Under a world trade agreement, Europe is now committed to drastic cuts in cereal production. There will be more 'set-aside', not less, and payments are to be even more generous. No more farmers will be paid to go right out of production but every one with more than 16 hectares of arable land will be compelled to put aside part of it,

either permanently or for one year.

The Conservatives defend this as a victory for Britain, because the rules are more flexible here than on the Continent. But Labour Treasury spokesman Nick Brown claims the Tories are too tied in with farming interests to put up a serious fight.

Farmers are now organising to defend themselves against the acusation that they are growing fat on subsidies.

John Cousins, who runs the Farming and Wildlife Advisory Group, says it is time to be 'positive about set-aside'.

He says: 'We have been pilloried for years about spray chemicals and nitrates, and now suddenly 15 per cent of farming land is not being sprayed and not being damaged. I set aside a substantial part of my farm, but I have done it to create margins around the hedgerows because wildlife likes those areas.'

Mr Gosling aso sees his career change, at the age of 42, as something positive. But in blunt economic terms, no amount of subsidy disguises what happened to him: he was made redundant.

More and more farming land is fated to go the same way. Over half a million hectares of British farmland, an area bigger than Lincolnshire, will lie fallow this year, at a cost of around £125 million. Someone, somewhere will have to decide what is to be done with it.

Figure 12.26 Watching the grass grow (*Source: The Observer*, 5 June 1994)

Figure 12.27 Opinions on set-aside

hectare, provided they set aside 15 per cent of the area claimed. In 1994 UK farmers were paid £253 per hectare for fallow, £140 for cereals and £445 for oilseed rape. That year, 700 000 hectares were set aside and farmers received payments totalling over £900 million.

Food shortages in the EC

One of the few agricultural products that the EC is short of is vegetable oil (Fig. 12.21). Oilseed rape is the only temperate crop that is likely to fill this gap.

—————— ? ——————

15 Study Figure 12.28.

a Use tracing paper to draw a map to show land use on the farm.

b Estimate the proportion of arable land that is set aside. Is the farmer complying with the rules of the set-aside policy?

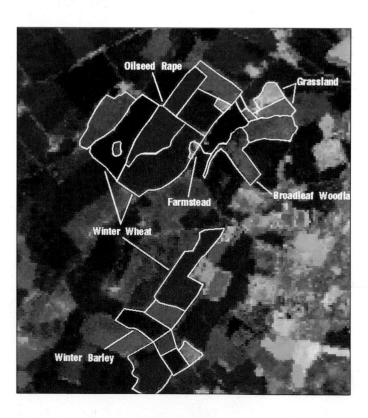

Figure 12.28 Landsat image of a Bedfordshire farm, UK, identifying farm boundaries and land use. Such images are analysed by the MAFF to check that what is growing in fields matches claims submitted by farmers. Farming fraud is estimated to cost the European taxpayer £2–6 billion a year.

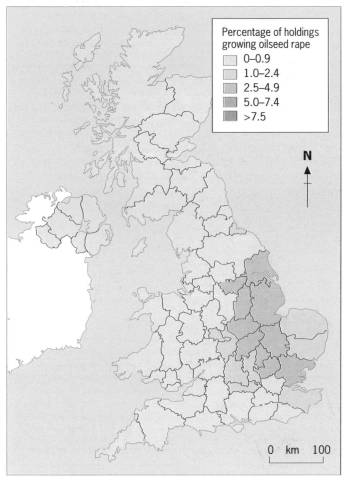

Figure 12.29 Oilseed rape in the UK, 1981

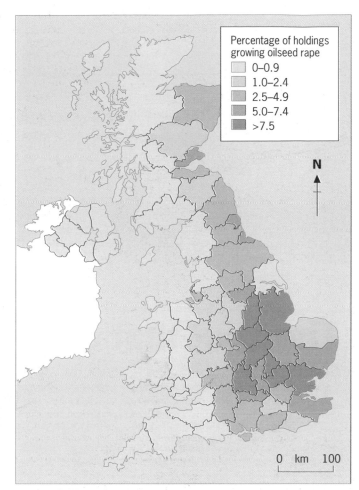

Figure 12.30 Oilseed rape in the UK, 1991

Oilseed rape in the UK

For many years the EC encouraged the cultivation of oil seed rape by paying the seed crushers. These, in turn, had to pay farmers a good price. The seed has a 40 per cent oil content and the residue is high in protein which is valuable for livestock feed. A further advantage is that oilseed improves the structure and fertility of the soil, and is often grown the year before wheat as a 'break' crop.

Oilseed rape was first adopted by a group of farmers in Hampshire in 1973. Improvements in the climatic tolerance and disease resistance of modern oilseed varieties have since caused a spread in the distribution of this crop (Figs 12.29 and 12.30). In England in 1991, 9.3 per cent of holdings grew oilseed (only wheat and barley were more popular as arable crops). This innovation and diffusion (see Sections 7.10 and 7.11) of oilseed (and more recently linseed) has been rapid because of wide publicity in the farming press about the crop's advantages (Fig. 12.31). It will be interesting to see if reductions in EC subsidies will send this success story into reverse in the 1990s.

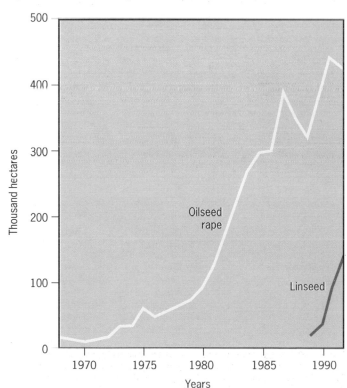

Figure 12.31 The success of oilseed rape and linseed in the UK, 1968–92

 CAP

?

16 Study Figures 12.29 and 12.30.
a How did the distribution of oilseed in the UK spread between 1981 and 1991?
b Suggest the factors (behavioural, physical, spatial) that might be responsible for the spread of oilseed in the UK.

17 Outline the main effects of subsidies on food surpluses and shortages in the EC.

18 The CAP is controversial and raises many issues. Look at Figures 12.25–12.27, 12.32–12.33 and research the latest developments of the CAP. You are a civil servant working in the Ministry of Agriculture, Fisheries and Food (MAFF) who has been asked to examine critically the workings of the CAP.
a Write a report for the Minister which sets out the impact of the CAP in the following areas: food prices, environment, agriculture in economically developing countries, prosperous/less prosperous farmers, and prosperous/less prosperous farming regions in the EC, governments, population.
b In your report briefly describe the recent attempts by the EC to reform the CAP.
c Suggest ways in which the CAP might be reformed to resolve some of the problems outlined in your report.

The CAP that does not fit

THE ESSENTIAL flaw in the CAP is that it only supports some Community farmers — the wrong ones, those who do not need support.

Modern agricultural technology means there is little difference between production costs of the most efficient farmers in Europe and those elsewhere. The cost of producing a tonne of wheat in East Anglia or the Beauce differs little from the cost of growing it in Kansas or Alberta.

The Community's problem is that it still sets its support prices at the high level theoretically needed to bolster the incomes of the less-than-efficient majority. But no matter how high the price level, it is seldom high enough to provide a decent income for the small and inefficient. It is simply mathematically impossible for it to do so. A French hill farmer marketing 10 beef cattle a year worth £600 each gains little if the EC price is increased by five per cent; a 2000 hectare wheat farmer would, however, increase his income by £10 000 as a result of such a price rise.

Despite steady undertones of discontent from representatives of the larger farmers in the EC, such as the British NFU, independent economists estimate that EC support prices are still 20 per cent higher than necessary to keep efficient farmers in business. And the most efficient have the largest share of production.

Typically, 80 per cent of cereals production is in the hands of large and efficient farmers, while approximately 320 000 out of the four million dairy farmers own more than half the EC dairy herd, producing three quarters of the total milk supply.

Each year farmers get fewer and farms get larger. Thus, while both the real gross income of the industry and average farm incomes fall, the income of the more efficient usually increases. This explains the paradox of continued increases in output in the face of falling agricultural income. It also explains why, over the CAP's lifetime, the gap between the incomes of large and efficient farmers and those of the small landholders has widened.

There is also the problem of the inexactitude of the CAP support system. Contrary to popular belief, the EC does not generally pay 'subsidies to farmers': it pays them to traders in food. The bulk of pay-outs from the Brussels farm fund – around 80 per cent of the current annual expenditure of £20 billion – are to butchers, grain traders, cold storers and food processors, who are paid to store, process and export the produce of Europe's farms. Indirectly these subsidised activities provide the floor price in the market for farmers' output.

Such a system does not, however, lend itself to the direction of payments to those who need price and income support. Rather the reverse: only big operators gain from such a policy.

The areas which fail to gain from this system are precisely those at the centre of the current unrest: remote 'less favoured areas' with small farms, poor soils, unfavourable climatic and communications conditions and with ageing and declining populations. In northern Europe the greatest concentration of farm holdings suffering these disadvantages is found in the centre and south of France.

The problems of French lamb producers have little to do with fluctuations of the pound against the franc, 'unfair subsidies to British sheep farmers', the import of frozen lamb from New Zealand, or even the dumping of East German beef and other meat on the EC market. It is a straightforward matter of efficiency. British and Irish shepherds can deliver lamb to the French market at prices between 10 and 20 per cent lower than their French colleagues because they are more efficient. They did so before EC subsidies and would continue to do so if subsidies were removed.

The decline of rural communities — 'desertification' as the French call it — cannot be solved by tinkering with the CAP. If the policy is to help those communities it has to be radically reformed.

At present, only about five per cent of the total spent by Brussels on agriculture goes to improve the conditions in these depressed rural areas.

While recent changes in the beef and sheep policies have been designed to provide more 'social' direct support, and EC Commission's new policies will gradually improve living and working conditions in these remote areas, the Community is still a long way from providing policies which will maintain and sustain rural communities in the future.

Not only does the CAP not fit present conditions and needs; in its present form it never will. A policy designed to support commodity markets in a period of food shortage and agricultural under-development (the 1950s) cannot cope with the dynamics of what has become a highly organised and technically advanced industry. In a broader sense and more importantly, the policy cannot cope with the social and environmental problems which Europe's rural areas now face.

Brian Gardner is a Brussels-based policy analyst.

Figure 12.32 Some of the problems of CAP (*Source: The Guardian*, 26 Oct. 1990)

Stunted growth

Black Africa has a high profile in Britain, thanks mainly to Band Aid spectaculars on behalf of the famine stricken regions. But there is much more to Africa's food problems than a mere lack of rain.

Zimbabwe, for example, with a two million tonne maize stockpile, has the potential to feed not only itself but many other African countries at a fraction of the total production, storage and transport cost which Europe incurs in getting its products to the famine areas.

But the Common Agricultural Policy, combined with the vast stockpile of farm products amassed by the United States, pose a greater menace than Africa's droughts to Zimbabwe's fertile and well-managed acres. Together they threaten to push 4 200 export-orientated commercial farmers – mainly whites – into bankruptcy, while nearly a million emergent black farmers could be thrown back on to primitive subsistence agriculture.

With the accession of the Mediterranean countries to the EEC, even tobacco – the crop which white large-scale farmers felt was most secure – risks taking a cruel hammering on world markets. There are fears that the Community might encourage the mass production of low-grade leaf which would then be dumped on world markets far below cost.

The director of Zimbabwe's Commercial Farmers Union, David Hasluck, says: 'The French, in particular, hold the view that as long as there is a food deficiency in Africa, the dumping of European food surpluses can be justified on humanitarian grounds'.

Botswana is one of 63 countries which get special EEC help – Brussels guarantees to buy 18 916 tonnes of Botswanan beef a year. Botswana's problem is over-generosity: cattle farmers have increased their grazing land to such an extent that it is encroaching on areas frequented by wild animals.

The country's entire ecology is threatened if EEC 'help' continues at its present level, although Botswana's cattle farmers suspect that the EEC is now using the ecology argument to try to restrict beef imports because of the urgent need to reduce the size of the beef 'mountain'.

Morocco regards itself as the African country most affected by the CAP now that Spain and Portugal have joined the Common Market. Aside from phosphates, agricultural exports are Morocco's chief revenue earner and more than half of these are absorbed by the EEC.

Figure 12.33 The impact of the CAP worldwide (*Source: The Times*, 26 Nov. 1986) © Times Newspapers Ltd 1986

Summary

- According to Malthus, food shortages result from rapid population growth leading to overpopulation. Unless population growth is controlled, famine, poverty and disease are inevitable.

- Today's 'neo-Malthusians' argue that the planet's resources are finite and that continued population growth will lead not only to food shortages and famine, but irreparable environmental damage.

- Technologists believe there are sufficient agricultural resources to feed adequately a world population of 10 billion or more. They argue that technological change and innovation will enable food production to keep up with population growth.

- Governments have a profound influence on agriculture and food production at national and international scales, and in economically developed and economically developing countries.

- In command economies like China, the state owns all farmland and the means of production. Farmers have less freedom of action than those in the West.

- In free-market economies like the EC, government influence is indirect, leaving farmers free to select the enterprises of their choice.

- China's agriculture policies have undergone radical change in the last 45 years. Judged by the impressive growth in food production in China since 1949, these policies have been successful.

- There is a common policy for agriculture in the EC. The policy subsidises and protects the farming industry.

- The EC's agricultural policy is costly and controversial. It has created problems of food surpluses and environmental damage. A range of new policies has been developed to tackle these problems. So far their success has been variable.

- The rapid spread of crops such as oilseed rape and linseed in the EC is a direct result of government subsidies.

Appendices

A1 Spearman rank correlation

Correlation is a statistical technique for measuring the relationship or degree of association between two variables. Geographers are often interested in how change in one variable is affected by change in another – i.e. we assume that one variable is a causal factor, bringing about change in another. Causal factors are designated *independent variables (x)*; non-causal factors are *dependent variables (y)*.

In order to measure the strength of relationships we use a *coefficient of correlation* which varies in value from –1.0 to +1.0. These extremes represent perfect correlations, while a value close to zero indicates the absence of any statistical correlation. A positive coefficient tells us that as *x* increases, *y* increases. A negative coefficient means that an increase in *x* causes a decrease in *y*, or vice versa. Perfect correlations are rare in geography because:

1 There are many random factors (e.g. human free will) which affect geographical phenomena.

2 Dependent variables are normally affected by several causal factors.

3 There is often some degree of error in sampling methods and measurements.

Where a relationship between variables exists, geographers are usually satisfied if this is demonstrated with a coefficient ranging between +0.7 and 1.0, and –0.7 and –1.0.

The Spearman rank correlation coefficient uses data measured on an ordinal or rank scale, and is particularly useful in surveys of attitudes or decision-making, where respondents are often asked to rank their preferences. The equation for calculating the coefficient is:

$$\text{Spearman rank correlation coefficient} \quad (r_s) = 1 - \left[\frac{6 \Sigma d^2}{n^3 - n} \right]$$

where d = difference in rank order of each pair of values
Σ = summation sign
and n = number of pairs of values.

Example

Infant mortality and food consumption in economically developing countries

	kcalories/head/day	Infant mortality per 1000	Rank			
	x	y	x	y	d	d²
Bangladesh	1925	108	8	3	5	25
Costa Rica	2782	17	3	10	7	49
Egypt	3213	57	1	7	6	36
Ethiopia	1658	122	9	2	7	49
Guyana	2495	48	4	8	4	16
India	2104	88	6	6	0	0
Mozambique	1632	130	10	1	9	81
Pakistan	2200	98	5	5	0	0
Sudan	1996	99	7	4	3	9
Tunisia	2964	44	2	9	7	49

$$n = 10$$
$$d^2 = 314$$
$$6\Sigma\,d^2 = 1884$$
$$n^3 - n = 990$$

Spearman rank correlation coefficient $\quad (r_s) = 1 - \left[\dfrac{6\Sigma d^2}{n^3 - n} \right]$

$$= 1 - \left[\frac{1884}{990} \right] = -0.90$$

The correlation coefficient of - 0.90 suggests a strong negative relationship between infant mortality and levels of food consumption. However, the correlation is far from perfect: this is because food consumption is only one of several independent variables which control infant mortality.

Because the correlation between infant mortality and food consumption could be due to chance it is necessary to establish its statistical significance using the *t*-distribution.

$$t = r_{s.} \sqrt{\frac{n - 2}{1 - r_s^2}}$$

n = number of pairs of data

degrees of freedom $(n - 2)$

$$t = -0.90 \sqrt{\frac{10 - 2}{1 - (-0.90^2)}} = -5.84$$

If we refer to *t*-tables we find that the critical value at the 0.01 level with 8 degrees of freedom is 3.36. As our *t*-value is greater than this critical value, we conclude that the correlation coefficient of 0.90 is statistically significant, and that there is a strong negative association between infant mortality and food consumption.

Critical values of Student's *t*-distribution

df	Two-tailed significance levels (one-tailed levels in parentheses)				
	0.10 (0.05)	0.05 (0.025)	0.02 (0.01)	0.01 (0.005)	0.001 (0.0005)
1	6.31	12.71	31.81	63.66	636.6
2	2.92	4.30	6.97	9.93	31.60
3	2.35	3.18	4.54	5.84	12.92
4	2.13	2.78	3.75	4.60	8.61
5	2.02	2.57	3.37	4.03	6.86
6	1.94	2.45	3.14	3.71	5.96
7	1.90	2.37	3.00	3.50	5.41
8	1.86	2.31	2.90	3.36	5.04
9	1.83	2.26	2.82	3.25	4.78
10	1.81	2.23	2.76	3.17	4.59
11	1.80	2.20	2.72	3.11	4.44
12	1.78	2.18	2.68	3.06	4.32
13	1.77	2.16	2.65	3.01	4.23
14	1.76	2.15	2.62	2.98	4.14
15	1.75	2.13	2.60	2.95	4.07
16	1.75	2.12	2.58	2.92	4.02
17	1.74	2.11	2.57	2.90	3.97
18	1.73	2.10	2.55	2.88	3.92
19	1.73	2.09	2.54	2.86	3.88
20	1.73	2.09	2.53	2.85	3.85
21	1.72	2.08	2.52	2.83	3.82
22	1.72	2.07	2.51	2.82	3.79
23	1.71	2.07	2.50	2.81	3.77
24	1.71	2.06	2.49	2.80	3.75
25	1.71	2.06	2.49	2.79	3.73

	Two-tailed significance levels (one-tailed levels in parentheses)				
	0.10 (0.05)	0.05 (0.025)	0.02 (0.01)	0.01 (0.005)	0.001 (0.0005)
26	1.71	2.06	2.48	2.78	3.71
27	1.70	2.05	2.47	2.77	3.69
28	1.70	2.05	2.47	2.76	3.67
29	1.70	2.05	2.46	2.76	3.66
30	1.70	2.04	2.46	2.75	3.65
40	1.68	2.02	2.42	2.70	3.55
60	1.67	2.00	2.39	2.66	3.46
over 60		approximates to the normal distribution			
z	1.64	1.96	2.33	2.58	3.29

A2 Chi-squared test

The chi-squared statistic tests whether the observed frequencies of a given phenomenon differ significantly from those hypothesised. Chi-squared is only applicable where the following conditions are met:

1   Data must be in frequencies.

2   Frequencies must be in absolute values, not percentages or proportions.

3   All expected frequencies should be at least 1.

Two-sample chi-squared

In this version of chi-squared, two samples of observations are compared with each other rather than with a hypothesised population. The null hypothesis suggests that the two samples were drawn from the same population, and therefore any differences are due to random variation.

The calculation of expected frequencies and degrees of freedom is shown below. This example analyses differences in nutrition in economically developing countries in 1961/63, 1967/69, 1979/81, 1988/90 and 2000. The null hypothesis (Ho) states that there is no difference between the samples other than that due to random sampling variations from a common population.

The expected values under the null hypothesis are provided by the marginal totals of rows and columns, so that:

Table 1   Number of economically developing countries at different average levels of nutrition (kcals/head)

kcals/head	1961/63	1967/69	1979/81	1988/90	2000 (estimate)	row total (r)
< 1900	23	7	8	8	4	50
1900–2500	66	74	54	56	42	292
> 2500	5	13	32	30	48	128
Column total (k)	94	94	94	94	94	total 470

$$\text{Expected cell frequency} = \frac{\text{row total} \times \text{column total}}{\text{grand total}}$$

Table 2   Calculation of expected frequencies

kcals/head	1961/63	1967/69	1979/81	1988/90	2000	row total (r)
< 1900	(50 × 94)/470	(50 × 94)/470	(50 × 94)/470	(50 × 94)/470	(50 × 94)/470	50
1900–2500	(292 × 94)/470	(292 × 94)/470	(292 × 94)/470	(292 × 94)/470	(292 × 94)/470	292
> 2500	(128 × 94)/470	(128 × 94)/470	(128 × 94)/470	(128 × 94)/470	(128 × 94)/470	128
Column total	94	94	94	94	94	total 470

The logic of this procedure is that the probability of a value falling in the first row/first column cell is equal to $(50/470) \times (94/470) = 0.021$.

The chi-squared statistic is calculated from the following formula:

$$\text{chi-squared} = \overset{r}{\Sigma}\overset{k}{\Sigma}\left[\frac{(O - E)^2}{E}\right]$$

where O = observed frequencies, E = expected frequencies and Σ = summation sign.

The double summation sign means that the additions take place over both rows (r) and columns (k):

$$\text{chi-squared} = 77.78$$

Degrees of freedom are determined by the product of the number of rows (r) and columns (k), each less one. Using the 0.01 significance level and 8 degrees of freedom, the critical value is 20.29 (see chi-squared statistical tables). As the chi-squared statistic (77.78) exceeds the critical value, the difference between the two samples cannot be explained by random variation, and the null hypothesis is rejected. Thus we can assume that levels of nutrition in economically developing countries have improved significantly since 1961/63.

Table 3 Critical values of the chi-square distribution

df	Significance level 0.10	0.05	0.01	0.005	0.001
1	2.71	3.84	6.64	7.88	10.83
2	4.60	5.99	9.21	10.60	13.82
3	6.25	7.82	11.34	12.84	16.27
4	7.78	9.49	13.28	14.86	18.46
5	9.24	11.07	15.09	16.75	20.52
6	10.64	12.59	16.81	18.55	22.46
7	12.02	14.07	18.48	20.28	24.32
8	13.36	15.51	20.29	21.96	26.12
9	14.68	16.92	21.67	23.59	27.88
10	15.99	18.31	23.21	25.19	29.59
11	17.28	19.68	24.72	26.76	31.26
12	18.55	21.03	26.22	28.30	32.91
13	19.81	22.36	27.69	30.82	34.55
14	21.06	23.68	29.14	31.32	36.12
15	22.31	25.00	30.58	32.80	37.70
16	23.54	26.30	32.00	34.27	39.29
17	24.77	27.59	33.41	35.72	40.75
18	25.99	28.87	34.80	37.16	42.31
19	27.20	30.14	36.19	38.58	43.82
20	28.41	31.41	37.57	40.00	45.32
21	29.62	32.67	38.93	41.40	46.80
22	30.81	33.92	40.29	42.80	48.27
23	32.01	35.17	41.64	44.18	49.73
24	33.20	36.42	42.98	45.56	51.18
25	34.38	37.65	44.31	46.93	52.62
26	35.56	35.88	45.64	48.29	54.05
27	36.74	40.11	46.96	49.65	55.48
28	37.92	41.34	48.28	50.99	56.89
29	39.09	42.56	49.59	52.34	58.30
30	40.26	43.77	50.89	53.67	59.70
40	51.81	55.76	63.69	66.77	73.40
50	63.17	67.51	76.15	79.49	86.66
60	74.40	79.08	88.38	91.95	99.61
70	85.53	90.53	100.43	104.22	112.32
80	96.58	101.88	112.33	116.32	124.84
90	105.57	113.15	124.12	128.30	137.21
100	118.50	124.34	135.81	140.17	149.45

A3 Project work

Before planning a project there are a number of important considerations:

• Farmers are busy, especially during the summer and autumn at the very time when students are available for fieldwork. Some do not welcome questions because they already have to fill in several official questionnaires. It is sensible therefore to make contact in advance to ask whether an interview is possible.

• Public rights of way are marked on Ordnance Survey maps. These are paths that you are entitled to walk along to observe the agriculture of an area. Do not stray from these paths, and you should avoid trespassing on private land.

• Follow the Country Code: guard against all risk of fire; fasten all gates; keep dogs under proper control; keep to the paths across farmland; avoid damaging fences, hedges and walls; leave no litter – take it home; safeguard water supplies; protect wildlife, wild plants and trees; go carefully on country roads; respect the life of the countryside.

Data collection and fieldwork

Because food systems encompass both rural and urban areas, they are spatially very extensive. For practical purposes sampling is therefore necessary in order to allow work to be carried out with a minimum use of time and other resources. This may be achieved by a total coverage of the farms in a small area or cluster or by selecting individual farmers in a random or systematic fashion.

There are other means of data collection. Postal questionnaires are possible, although response rates are usually low. Telephone enquiries will save travelling, but again refusals will be higher than for face-to-face interviews.

Unfortunately there is no comprehensive list of farms available. The Ministry of Agriculture regards their records as confidential, as do the National Farmers' Union. The simplest approach is probably to use the Yellow Pages.

Besides asking farmers the obvious questions about their crops and livestock, it may be appropriate to elicit information about some of the following:

• The effect of the environment upon the farming system.

• The contributions of labour and capital. The best way to approach this is to find out what changes have taken place in employment patterns and capital investments over a period of time.

• The influence of proximity to markets and the impact of urbanisation.

• Decision-making by farmers.

• The impact of government and EC policies.

Remember that your respondents' time will be limited. The order of questions, manner of questioning and timing are important. Leading questions will distort the results, and blunt enquiries about age and income may cause offence.

Fieldwork by observation is an alternative to questionnaires. The Land Utilisation surveys of the 1930s and 1960s have been published in map form and provide a good basis for comparative analysis. One research question might be how much the various land use categories in your area have changed in the periods 1930s–1960s, and 1960s–1990s. These surveys were descriptive and tell us little about the factors behind land-use decision-making, so care must be exercised in interpretation.

Tracing food through marketing channels is a difficult and complex task. Visits to food processing factories are likely to be enlightening but food retailers may be unable or unwilling to reveal information. Many firms guard their own business secrets carefully.

Other food issues might also be the subject of survey work. Beginning with friends and relations, and then possibly extending the enquiry to other households, it would be especially interesting to ask about people's food habits. Do they differ according to income, size or family, and age as one might expect?

Be sure to include questions which allow the analysis of the data by the sort of variables that might classify responses (age, sex, income, education, neighbourhood). Asking questions about such sensitive topics is by no means easy, so think carefully about presentation.

A4 Useful publications

Central Statistical Office (1993) *Aspects of Britain: agriculture, fisheries and forestry* London: HMSO.
A basic background to agriculture in Britain, using the latest data.

Ministry of Agriculture, Fisheries and Food (Annual) *The digest of agricultural census statistics: United Kingdom* London: HMSO.
This is an especially useful publication. It gives data for counties and countries.

Ministry of Agriculture, Fisheries and Food (Annual) *Agriculture in the United Kingdom* London: HMSO.
This report is an overall description of trends in UK agriculture.

Ministry of Agriculture, Fisheries and Food (Annual) *Farm incomes in the United Kingdom* London: HMSO.
A more specialised collection of financial data.

Ministry of Agriculture, Fisheries and Food (Annual) *Household food consumption and expenditure: annual report of the National Food Survey Committee* London: HMSO.
This report describes the state of the nation's nutrition. There are sections on patterns over the previous few years and regional differences of food intake. There are also breakdowns by income group and household composition.

Ministry of Agriculture, Fisheries and Food (Weekly) *Agricultural market report (livestock and horticultural produce),* London.

Ministry of Agriculture, Fisheries and Food (Annual) *Basic horticultural statistics for the United Kingdom,* London.

Ministry of Agriculture, Fisheries and Food (Annual) *Statistics from the June agricultural and horticultural census and the December glass house census, England and Wales,* London.

Ministry of Agriculture, Fisheries and Food (Monthly) *Report on fruit and vegetable crops in England and Wales,* London.

Welsh Office (Annual) *Welsh agricultural statistics* Cardiff: Welsh Office.
Statistics relating to Wales are available from: Publication Unit, E & SS Division, Welsh Office, Cathays Park, Cardiff CF1 3NQ.

Welsh Office (Annual) *Farm incomes in Wales* Cardiff: Welsh Office.

Scottish Office, Agriculture and Fisheries Department (Annual) *Farm incomes in Scotland* Edinburgh: Scottish Office.
Statistics on Scotland can be obtained from: Scottish Office Library, Publication Sales, Room 1/44, New St Andrew's House, Edinburgh EH1 3TG.

Scottish Office, Agriculture and Fisheries Department (Annual) *Agriculture in Scotland* Edinburgh: Scottish Office.

Scottish Office, Agriculture and Fisheries Department (Annual) *Economic report on Scottish agriculture* Edinburgh: Scottish Office.

Department of Agriculture, Northern Ireland (Annual) *Statistical review of Northern Ireland agriculture* Belfast: DANI. Northern Ireland data are published by: DANI., Dundonald House, Upper Newtonards Road, Belfast BT4 3SB.

Department of Agriculture, Northern Ireland (Annual) *Farm business data (Northern Ireland)* Belfast: DANI.

Department of Agriculture, Northern Ireland (Annual) *Agricultural market report (Northern Ireland)* Belfast: DANI.

Department of Agriculture, Northern Ireland (Annual) *Horticultural market report (Northern Ireland)* Belfast: DANI.

Department of Agriculture, Northern Ireland (Annual) *Northern Ireland agricultural census statistics* Belfast: DANI.

Commission of the European Communities (Annual) *The agricultural situation in the Community.* There are many series published by the EC but this is the one with the most useful, basic data.

Food and Agriculture Organisation of the United Nations (Annual) *Production yearbook* Rome: FAO.

Food and Agriculture Organisation of the United Nations (Annual) *The state of food and agriculture* Rome: FAO.

The Ministry of Agriculture, Fisheries and Food publish agricultural maps. Care should be taken in their use, however, because many were compiled in the 1960s and 1970s. Contact MAFF Publications, London SE99 7TP.

Agricultural Land Classification Maps: 190 sheets at 1:63 360 scale, with accompanying reports.

Agricultural Land Classification Maps: 6 sheets at 1:250 000 scale.

Agricultural Land Classification Map: 1 sheet at 1:625 000 scale.

Type of Farm Maps: 8 sheets at 1:250 000 scale, in two series: classification by acreage or Standard Man Days.

Type of Farm Map: 1 sheet at 1:625 000 scale.

Land Utilisation Survey: sheets covering the whole of Britain at 1:63 360.
The survey was undertaken 1931–41. Available in some reference libraries and university map libraries.

Second Land Utilisation Survey: partial coverage of England and Wales at 1:25 000.
The surveys date from the early 1960s.

References

Alamgir, M (1980) *Famine in south Asia: political economy of mass starvation* Cambridge, Mass.: Oelgeschlager, Gunn and Hain.

Alexandratos, N (ed.) (1988) *World agriculture: toward 2000. An FAO study* London: Belhaven.

Atkins, P J (1977) The intra-urban milk supply of London, circa 1790–1914, *Transactions of the Institute of British Geographers* new series 2, 383–99.

Atkins, P J (1987) The Charmed Circle: von Thünen and agriculture around nineteenth century London, *Geography* 72, 129–39.

Barrow, C J (1991) *Land degradation: development and breakdown of terrestrial environments* Cambridge: Cambridge University Press.

Barrow, C J (1987) *Water resources and agricultural development in the Tropics*, Harlow: Longman.

Barry R G and Chorley, R J (1968) *Atmosphere, weather and climate* London: Methuen.

Bibby J S and Mackney D (1969) 'Land use capability classification' *Technical Monograph 1, Soil survey of Great Britain*.

Blaxter K L and Fowden L, (1982) *Food nutrition and climate* Oxford, Applied Science Publications.

Bongaarts, J (March 1994) 'Can the growing human population feed itself?', *Scientific American*.

Bowden, L W (1965) Diffusion of the decision to irrigate, *University of Chicago Department of Geography Research Paper* No. 97.

Bowler, I R (ed.) (1992) *The geography of agriculture in developed market economies* Harlow: Longman.

Bridges, E M (1978) *World Soils*, 2nd edn, Cambridge: CUP.

Buringh, P and Dudal, R (1987) Agricultural land use in space and time, pp 9–43 in Wolman, M G and Fournier, F G A (eds) *Land transformation in agriculture* Chichester: Wiley.

Central Statistical Office (1993) *Annual abstract of statistics* London: HMSO.

Dixon, C (1990) *Rural development in the third world* London: Routledge.

European Commission (1993) *The agricultural situation in the community: 1992 report* Luxembourg: EC.

The Economist (30 April 1994), 'Famine in the Horn of Africa' London: Economist Publications.

Fakhfakh, M (1979) *Atlas de Tunisie*, Paris: Éditions Jeune Africque.

Fearne, A P (1990) Communications in agriculture: results of a farmer survey, *Journal of Agricultural Economics* 41, 371–80.

Foster, P (1992) *The world food problem* Boulder: Rienner.

Gasson, R (1992) Farmers' wives: their contribution to the farm business, *Journal of Agricultural Economics* 43, 74–87.

Grigg, D B (1984) *An introduction to agricultural geography* London: Hutchinson.

Grigg, D B (1992) World agriculture: production and productivity in the late 1980s, *Geography* 77, 97–108.

Hall, P G (ed.) (1966) *Von Thünen's isolated state* Oxford: Pergamon.

Ilbery, B W (1985) *Agricultural geography: a social and economic analysis* Oxford: Oxford University Press.

Ilbery, B W (1992) *Agricultural change in Great Britain* Oxford: Oxford University Press.

Jazairy, Alamgir, M and Panuccio, T (1992) *The state of world rural poverty* London: Intermediate Technology Publications.

Kohls, R L (1967) *Marketing of agricultural products* 3rd edition New York: Macmillan.

Leach, G (1976) *Energy and food production* Guildford: IPC.

MacCarty, H H and Lindberg, J B (1966) *A preface to economic geography* New Jersey: Prentice-Hall.

Mackney D and Burnham, C P (1964) 'The soils of the West Midlands' *soil survey of Great Britain, England and Wales Bulletin No. 2*.

Marks, H F (ed.) (1992) *Food: its production, marketing and consumption* Reading: University of Reading.

Marland, A (1990) 'An overview of organic farming in the United Kingdom' *Outlook on Agriculture 18*, 24–27.

Ministry of Agriculture, Fisheries and Food (1990) *Household food consumption and expenditure 1989* London: HMSO.

Ministry of Agriculture, Fisheries and Food (1991) *Agriculture in the United Kingdom: 1990* London: HMSO.

Ministry of Agriculture, Fisheries and Food (1993a) *Agriculture in the United Kingdom, 1992* London: HMSO.

Ministry of Agriculture, Fisheries and Food (1993b) *Farm incomes in the United Kingdom, 1991/92 edition* London: HMSO.

Marland, A (1990) An overview of organic farming in the United Kingdom, *Outlook on Agriculture 18*, 24–27.

Morgan, W T W M (1988) Tamilnadu and eastern Tanzania: comparative regional geography and the historical development process, *Geographical Journal* 154, 69–86.

Organic Farmers and Growers Ltd. (unpublished data)

Oxfam in association with the *Observer* (1988) Fragile future, © Paul Harrison.

Pannell, C W (June 1982) 'Croplands to feed China' *Geographical Magazine* LIV, No.6.

Pearce, D, Barbier, E and Markandya, A (1990) *Sustainable development* London: Earthscan.

Poleman, T T (1981) *Quantifying the nutrition situation in developing countries'* Stanford University, food research institute studies 18, 1 58.

Potato Marketing Board *Usage and attitude survey of the British potato market 1991/2.*

Saarinen, T F (1966) 'Perception of drought hazard on the Great Plains' *University of Chicago, Dept of Geography, research paper 106.*

J Sainsbury's (1991) *Study pack: information* London: Sainsbury's.

J Sainsbury's (1992) *Study pack: marketing and merchandising* London: Sainsbury's.

Selby, J A (1987) On the operationalisation of Pred's behavioural matrix, *Geografiska Annaler* 69B, 81–90.

Sen, A (1981) *Poverty and famines* Oxford: Oxford University Press.

Simmons, I G (1987) Transformation of the land in pre-industrial time, pp 45–77 in Wolman, M G and Fournier, F G A (eds) *Land transformation in agriculture* Chichester: Wiley.

Simon, H A (1975) *Models of man: social and rational,* New York: Wiley.

Sinclair, R J (1967) 'Von Thünen and urban sprawl' *Annals of the Association of American Geographers 57,* 72–87.

Spedding, C R W, Walsingham, J M and Hoxey, A M (1981) *Biological efficiency in agriculture* London: Academic Press.

Spedding, C R W (1988) *An introduction to agricultural systems* 2nd edition London: Elsevier.

Tabor, S (1979) *Notes on Brazilian consumption and expenditure survey* Washington DC, US Department of Agriculture, Economics, Statistics, and Cooperatives Service.

Thouvenot, C (1987) *Le pain d'autrefois: chroniques alimentaires d'un monde qui s'en va* Nancy: Presses Universitaires de Nancy.

Tivy, J (1990) *Agricultural ecology* Harlow: Longman.

UNICEF (1992) *The state of the world's children* Oxford: Oxford University Press.

Unilever (1993) *Annual Review 1993.*

Whynne Hammond, C (1985) *Elements of human geography,* 2nd edition London: Unwin Hyman.

Wolpert, J (1964) The decision-making process in a spatial context, *Annals of the Association of American Geographers* 54, 537–58.

Glossary

Absentee landlords Land owners who live away from their estate, often in an urban area. They take rent from tenant farmers but may make little contribution in return.

Accumulated temperatures For a specified period, a total in degrees celcius or fahrenheit of the mean daily temperatures above a certain lower limit. From this it is possible to see whether a certain crop can be grown in that location.

Adret A slope which faces towards the equator and is therefore exposed to the sun's rays.

Agribusiness Large-scale capital-intensive farming which is sometimes linked to other parts of the food system, especially processing factories and supermarkets.

Agricultural regions Areas of farming which share similar characteristics of production.

Agrochemicals Chemicals used on the farm, such as fertilisers, pesticides, herbicides and insecticides.

Agro-ecosystems Ecosystems which have been modified by people for agricultural purposes.

Alternative farming Agriculture which is different from the intensive form of farming which relies on oil for fuel and agrochemicals. Alternatives include the more environmentally friendly 'organic' production.

Annuals Plants which are replaced each year by the farmer.

Aspect The direction in which a slope faces, which often affects the amount of incoming solar radiation.

Biodiversity The number and variety of plant and animal species found in an ecosystem.

Biomass The mass of biological material present in a given area.

Biomes Continental scale, zonal ecosystems such as temperate deciduous forest, tropical rain forest, tundra, and desert.

Brown earth Brown soils formed from a wide variety of parent materials by moderate leaching under deciduous forest.

Capitalism/capitalist A system of economic organisation based upon the exchange of commodities for money, in which the forces of the market are more important than government planning.

Carnivore (top carnivore) A meat eating animal.

Carrying capacity The maximum number of people who can be sustained by the resources of a particular area.

Centrally planned economies Socialist countries where much of the decision-making about the nature and structure of the food system is controlled by the government.

Chestnut soils Brown soils of the grasslands in the Russian steppes and North American prairies.

Chernozems Black, crumbly soils which support productive arable farming. They are found extensively in Russia.

Collectives A form of socialist agriculture in which farmers join together to provide labour for their common benefit and to meet production targets set by the state.

Command economy An economic system in which production is controlled by the government through the ownership and control of farms, factories and shops.

Commercial agriculture Farming motivated by profit, where food is produced on large-scale holdings by advanced technological means.

Contour ploughing The practice of ploughing along the contours of a slope in order to minimize the downslope run-off of water and thereby prevent soil erosion.

Co-operatives A group of farmers with a shared interest such as the bulk purchase of inputs, the sharing of machinery or the joint negotiation of sales contracts.

Crop combination regions A form of agricultural regionalisation which takes account of the variety of crops grown in each area. The combinations of crops (and livestock) give the region its distinctive character.

Crop rotation A method of farming which avoids growing the same crop in a field continuously. A succession of crops maintains soil fertility and reduces the risk of pests and diseases.

Cultivars Plant species which have been made useful to humans by selective breeding.

Cultural landscapes Natural landscapes as modified by different cultural groups.

Degree days For a given period, the difference between the mean daily temperatures and a given lower threshold.

Desertification The reduction in agricultural capacity as a result of human and/or natural processes. Full desert-like conditions are rare but productivity and population density are greatly reduced. Factors may include overgrazing, deforestation, desiccation due to climate change, etc.

Downstream functions The stages through which food passes after leaving the farm, e.g. bulking, processing, packing, storage, wholesaling, retailing.

Economic wo/man The decision-maker in classical economic theory who has perfect knowledge and aims to get the most for the least, for instance by achieving either maximum profits or minimum costs.

Economies of scale The cost savings gained by production on a large scale.

Ecosystem A community of plants and animals which interact with each other and with their physical environment.

Enterprises The divisions of farming, e.g. cereal-growing or milk production, in which farmers can specialise, or combine, in a mixed farming region.

Entitlements The purchasing and bargaining power which gives people access to food and other basic needs, either as cash or kind earned through wage labour, the sale of assets, or the exercise of power gained from occupying a privileged place in society.

Evaporation The changing state of water from liquid to vapour by diffusion into the atmosphere from an exposed water surface.

Exchange entitlements Entitlements to food gained by market mechanisms such as the use of cash earned through wage labour or the sale of assets.

Extensification/Extensive The process of curbing the productivity of the land by reducing investment in inputs such as machines and agrochemicals e.g. hill farming.

Field capacity The amount of water which a body of soil will hold in its voids and pores after excess moisture has drained away.

Food availability decline (FAD) A decrease in food supply which may lead to undernutrition and famine.

Food chain The succession of levels through which energy is transferred from herbivores to carnivores, and on to top carnivores, by eating and being eaten.

Food security When there is access to sufficient food for individuals to lead a healthy life.

Food system The chain which facilitates the rapid and efficient movement of food from the farmer's field to the dinner table. It includes a number of distinct actors and operations, notably: producers, processors, wholesalers, retailers, and consumers.

Frost hollows A topographic depression, such as a valley bottom, which allows the accumulation of cold, dense air, causing frosts. A notorious frost hollow, which has experienced ground frosts in mid summer, is at Houghall near Durham city.

Genetic manipulation A form of biotechnology which involves the modification of a plant or animal's genetic make-up in order to introduce new, desirable characteristics such as enhanced pest resistance.

Global warming The process by which the earth's atmosphere is gradually increasing its mean temperature. This is usually thought to be the result of fossil fuel combustion over the last 200 years.

Green belt An area adjacent to a built-up area where further development is restricted by a planning authority.

Green revolution The attempt to improve the productivity of crops in developing countries which began in the 1960s with the breeding of new high-yielding varieties of wheat and rice.

Greenhouse effect The property of the gaseous mix in the atmosphere which allows the sun's energy to be absorbed and held. It is thought that the excessive release carbon dioxide and other gases into the atmosphere is upsetting its delicate balance as more heat is retained, causing global warming.

Herbivores Plant eating animals such as cattle and sheep.

Humus Decomposed organic matter which accumulates in and within the soil.

Hunter-gatherer Humans who make a living from hunting animals and collecting edible fruits, seeds, roots, tubers and leaves (e.g. bushmen of the Kalahari, aborigines of Australia).

Hydrological growing season The mean number of days in a year when there is sufficient water to allow plants to grow. The term is usually applied in the tropics where the thermal growing season covers the whole year.

Inorganic An object/substance of non-biological origin e.g. a chemical fertiliser.

Inputs The investments necessary on a farm to produce food. They include labour, machinery, seeds, fertiliser, fodder, and pesticides.

Inselbergs Isolated hills with steep sides.

Intensification/Intensive The process whereby farmers increase the productivity of their land through investment in more inputs, such as machines and agrochemicals.

Interfluves The intersections between neighbouring river valleys.

Land degradation The deterioration of land suitability for agriculture by processes such as soil erosion, desertification, and salinisation.

Leach The downward movement of water which washes out soluble nutrients and other ingredients from the upper levels of a soil profile by the action of rain or irrigation.

Legumes Plants which have the ability to fix nitrogen in the soil, thus enhancing fertility. The pods of legumes are called pulses, some of which are edible and high in vegetable protein, such as peas and beans.

Ley A period of years during which grass is grown in a field as part of an arable rotation.

Locational rent The difference between the value of a farm's output and the joint costs of production and transport to market.

Loess A soil which is porous and generally fertile; originally a wind deposit of fine particles of silt or dust.

Marginal return The rate of change of monetary reward from farming as input costs vary.

Marketing channels Systems of food distribution. These vary from a one-stage channel, with direct contact between the farmer and consumer, to multi-stage channels which might involve farmer, manufacturer, wholesaler and retailer.

Monoculture An agricultural system in which a single crop is grown continuously in the same field.

Negative feedback A process which maintains the equilibrium of a system.

Neighbourhood effect An effect upon people's views, decisions and actions caused by the influence of living in a particular location. The effect is localised because of the friction of distance.

Nomadic pastoralist Livestock farmers who are mobile for at least part of the year, usually in search of water and grazing for their animals.

Off-land enterprises Highly intensive farms which are so restricted in area that they may resemble factories more than the traditional image of a farm with green meadows and fields of corn. Livestock such as laying fowl and broilers, pigs, and sometimes milch cattle are kept this way.

Omnivore An animal which eats both plants and other animals e.g. human being.

Organic An object/substance of plant or animal origin. Organic farming avoids the use of synthetic/inorganic chemical fertilisers, herbicides or pesticides.

Optimisers In economic terms, a decision-maker whose goal is either to maximise profits or minimise costs.

Optimum temperatures The best thermal conditions for crop growth, not too hot nor too cold.

Outputs The products of agriculture, especially crops and livestock for food and fibre.

Over-cultivation The excessive exploitation of land to the point where productivity falls due to soil exhaustion or land degradation.

Over-population A surplus of population beyond the carrying capacity limits of local resources.

Peasant agriculture Small-scale farming in economically developing countries, in which subsistence still plays a part.

Peds Aggregates of soil particles which make up the soil's structure and influence its characteristics, such as water-holding capacity, permeability, aeration and heat transfer.

Perception Decision-makers such as farmers may operate on the basis of their view of the world, not according to objective reality. Their perceptions are coloured by cultural biases, their own individual personalities, and by the quantity and quality of information that is available to them.

Perfect knowledge The theoretical ability to have full information about costs, prices and market opportunities in order to make fully rational locational or production decisions.

Perennials Plants which may yield a crop for several seasons, for instance fruit trees or vines.

Photosynthesis The process by which sunlight is converted into energy for plant growth.

Podsols Soils which form in cool, wet conditions where leaching carries nutrients to lower levels. Podsols are often infertile, with heathland vegetation or coniferous forest.

Positive feedback A virtuous circle in a system which mutually reinforces positive change.

Precipitation effectiveness The amount of water available to plants after a deduction for losses due to run-off, evaporation and transpiration.

Precipitation The deposition of water from rain, snow, hail, dew and frost.

Primary industry Production of raw materials by farms, forests, fisheries and mines.

Processor That part of a system which changes the inputs into the outputs.

Producer The first member of the food system who is responsible for raw products such as unmilled grain or wet fish.

Redistributive agriculture The dominant farming type in socialist countries, in which collective and state farms produce food according to targets set by the state.

Regionalisation The process of dividing the earth's surface into areal units such as agricultural regions.

Ridge-and-furrow A distinctive ploughing system, common in medieval England, in which long parallel ridges of earth were created.

Rural-urban fringe The transition zone between the continuously built-up urban area and the open countryside.

Salinisation The accumulation in the soil of salts which are toxic for plants, often to the point where agriculture cannot continue.

Satisficers Decision-makers who are satisfied with *less* than a goal such a maximum profitability.

Scale economies See 'economies of scale'.

Seasonal distribution The monthly variations of precipitation, which can have a significant effect upon plant growth.

Secondary industry The processing and manufacturing sector.

Set-aside The land on which a farmer is required by the EC to cease production of surplus commodities such as wheat.

Share tenants Farmers who do not own the land they cultivate. Their rent (in cash or kind) is a fixed percentage of the harvest each year.

Shifting cultivators Farmers who clear and burn an area of vegetation and then grow crops for a few years until yields decline. They then move on to a new clearing.

Soil erosion The loss of soil from a field's surface by the action of wind or water. Certain types of soil are susceptible to erosion, as are soils on slopes. Erosion can be reduced by soil conservation measures which may involve a change in cultivation practices.

Soil moisture budget The balance in a body of soil between inputs of moisture from infiltration and outputs from sub-surface gravity flow, evaporation, and extraction by plants.

Soil structure The way individual particles of soil bond together. Examples are prismatic, columnar, angular blocky, platy, and granular.

Soil texture The particle size composition of a soil.

State farms A socialist form of farming in which the state makes all production decisions and the farmers are merely paid workers.

Store A component of a system in which energy and matter are held.

Subsistence agriculture Farming in which the production of food for household consumption is the priority. Total self-sufficiency is now rare.

Sustainable agriculture   Farming which meets present demand without reducing the ability of future generations to achieve their needs e.g. farming which avoids pollution and soil erosion.

System   Systems thinking emphasises the connections and relationships between a group of elements, often called a system.

Temperate crops   Economic plants which are best suited to the climate of cooler, temperate latitudes.

Thermal growing season   The number of days in a year when the mean daily temperature reaches a level sufficient for plant growth.

Tissue culture   The microscopic subdivision of plant material to enable the growth of many new plants.

Transpiration   The loss of water from the surface of a plant's leaves.

Ubac   A slope which faces away from the equator and may therefore be shaded from the sun's rays.

Variance   A measure of variability based on squaring deviations between observed and expected values.

Workability   The ease with which soil can be cultivated. This is affected by soil properties. Thus clay soils may be difficult to work because they are sticky when wet and hard when dry.

Index

Published by Collins Educational
77–85 Fulham Palace Road
London W6 8JB

An imprint of HarperCollins Publishers

© 1995 Michael Raw and Peter Atkins

First published 1995

ISBN 0 00 3266664

Michael Raw and Peter Atkins hereby assert their moral right to be identified as the Authors of this Work.

Edited by Ela Ginalska and Anne Montefiore
Designed by Jacky Wedgwood
Picture research by Caroline Thompson
Computer artwork by Barking Dog Art, Contour Publishing, Jerry Fowler
Illustrations by Joan Corlass, Jeremy Gower, Dave Parkin

Typeset by Create Publishing Services Ltd, Bath, Avon

Printed and bound by Rotolito Italy

Dedications

MDR: To my parents
PJA: To Liz, Rowan and Miranda

The authors wish to acknowledge the following for proffering advice, encouragement and materials for the Work: Michael Alexander, David Cowton, Nicholas Cox and Dai Morgan (Department of Geography, University of Durham), David Bennison (Department of Retailing and Marketing, The Manchester Metropolitan University), Bridget Williams (Sainsbury's), Jill Gatcum (Unilever) and Mr J and Mrs K G Atkins.

The authors and publisher are grateful to the following for their comments on the manuscript: Dr John Alliston, Royal College of Agriculture; Mike Battcock, Intermediate Technology; Alison Evennett, British Nutrition Foundation; Dr P Jones, Climatic Research Unit, University of East Anglia; Prakash Shetty and Jeremy Shoham, Human Nutrition Centre, London School of Hygiene and Tropical Medicine.

Acknowledgements

Every effort has been made to contact the holders of copyright material, but if any have been inadvertently overlooked the publisher will be pleased to make the necessary arrangements at the first opportunity.

Photographs
The publisher would like to thank the following for permission to reproduce photographs:

Aerofilms, Figs 2.16, 8.7;
Michael Alexander, Fig. 5.10;
Peter Atkins, Figs 2.3, 2.5, 2.10, 4.34, 11.14;
Barnaby's Picture Library, Figs 9.6, 9.14;
Anthony Blake Photo Library/Bernkastel, Fig. 4.17;
British Library, Fig. 6.1;
The J Allan Cash Photolibrary, Figs 4.16, 10.2, 11.11;
© Crown Copyright/MOD - reproduced with permission of the controller of HMSO, Fig. 6.5;
Empics/Ian Crawford, Fig. 11.23;
Mary Evans Picture Library/Alexander Meledin Collection, Fig. 10.18;
Mary Evans Picture Library, Fig. 10.23;
Julia Waterlow/Eye Ubiquitous, Fig. 6.10;
GeoScience Features Picture Library, Fig. 5.9;
Adam Woolfitt/Robert Harding Picture Library, Figs 2.19, 6.3, 8.15;
Jim Holmes, Fig. 7.27;
Holt Studios International, Figs 8.1, 11.16, 12.7, 12.8, 12.9, 12.11;
The Hutchison Library, Figs 7.10, 9.2, 11.2, 12.13;
The Illustrated London News Picture Library, Fig 10.21;
Jennifer Lucas, Fig. 11.21;
McDonald's Restaurants Ltd, Fig. 9.24;
David Mansell, Fig. 12.26;
NHPA, Figs 2.4, 2.17, 3.7, 4.8, 6.24, 11.3;
Mike Goldwater/Network, Fig. 3.2;
Barry Lewis/Network, Fig. 10.8;
The Robert Opie Collection, Fig. 8.16;
Oxford Scientific Films Ltd, Fig. 5.23;
Panos Pictures, Figs 1.8, 3.3, 6.2, 6.4, 6.11, 6.14, 7.15, 10.26, 11.7, 12.14;
Range Pictures Ltd, Fig. 10.27;
Michael Raw, Figs 1.2, 4.1, 5.21;
J Sainsbury plc, Figs 9.15, 9.18;
Science Photo Library, Figs 4.14, 12.28;
SCOPE, Fig. 6.15;
Mark Edwards/Still Pictures, Figs 3.8, 4.2, 5.18;
Tony Stone Images, Figs 2.12, 3.6, 4.9, 4.39, 5.20, 9.7, 11.2;
C & S Thompson, Figs 5.11, 8.17;
Unilever, Fig. 8.24;
Tony Waltham Geophotos, Fig. 8.21;
Simon Warner, Figs 3.5, 7.6;
David Woodfall, Figs 2.18, 12.10;
ZEFA, Figs 7.28, 12.18.

Cover picture
Rice terraces in the Philippines. Source: ZEFA

Maps
Fig 4.15, reproduced with the permission of the Controller of Her Majesty's Stationery Office, © Crown Copyright, from: Wensleydale and Upper Wharfedale 1989 1:50000 Ordnance Survey Map.
Fig. 5.14, map extracted from Tadcaster Sheet Second Land Utilisation Survey of Britain, Director Alice Coleman.